T0258045

Green Chemistry

Green Chemistry

Edited by **Ralph Britton**

New York

Published by NY Research Press,
23 West, 55th Street, Suite 816,
New York, NY 10019, USA
www.nyresearchpress.com

Green Chemistry
Edited by Ralph Britton

International Standard Book Number: 978-1-63238-215-3 (Hardback)

Printed in the United States of America.

Contents

Permissions

List of Contributors

Preface

Over the recent decade, advancements and applications have progressed exponentially. This has led to the increased interest in this field and projects are being conducted to enhance knowledge. The main objective of this book is to present some of the critical challenges and provide insights into possible solutions. This book will answer the varied questions that arise in the field and also provide an increased scope for furthering studies.

This book gives a comprehensive discourse on the various aspects of handling various environmental concerns through the use of various approaches involving chemistry. Green chemistry can also be called environmental chemistry. In essence, it is philosophical approach which propagates a viewpoint of applying chemistry in creating more environment friendly solutions. It is deemed to be a critical part of an extensive plan for protection of human well-being and ecology. It is an economically driven and scientific methodology towards accomplishing the goals of ecological protection and sustainable progress. Incorporating technological advancements with safety of the environment is one of the crucial challenges of current times. This book explains unique approaches for the industrial processes as well as chemical laboratories which are gentle to the environment. To present an overview on step change technologies, this book has been compiled from contributions by green chemistry experts.

I hope that this book, with its visionary approach, will be a valuable addition and will promote interest among readers. Each of the authors has provided their extraordinary competence in their specific fields by providing different perspectives as they come from diverse nations and regions. I thank them for their contributions.

Editor

Green Chemistry –
Aspects for the Knoevenagel Reaction

Ricardo Menegatti
Universidade Federal de Goiás
Brazil

1. Introduction

Knoevenagel condensation is a classic C-C bond formation reaction in organic chemistry (Laue & Plagens, 2005). These condensations occur between aldehydes or ketones and active methylene compounds with ammonia or another amine as a catalyst in organic solvents (Knoevenagel, 1894). The Knoevenagel reaction is considered to be a modification of the aldol reaction; the main difference between these approaches is the higher acidity of the active methylene hydrogen when compared to an α-carbonyl hydrogen (Smith & March, 2001).

Figure 1 illustrates the condensation of a ketone (1) with a malonate compound (2) to form the Knoevenagel condensation product (3), which is then used to form the α,β-unsaturated carboxylic compounds (3) and (4) (Laue & Plagens, 2005).

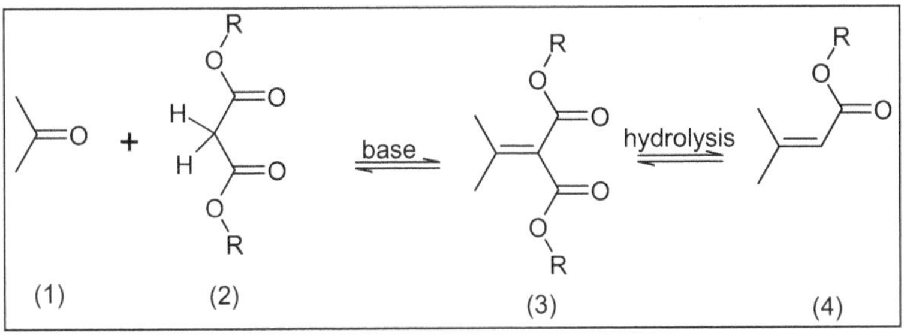

Fig. 1. An example of the Knoevenagel reaction.

Subsequent to the first description of the Knoevenagel reaction, changes were introduced using pyridine as the solvent and piperidine as the catalyst, which was named the Doebner Modification (Doebner, 1900). The Henry reaction is another variation of the Knoevenagel condensation that utilises compounds with an α-nitro active methylene (Henry, 1895). The general mechanism for the Knoevenagel reaction, which involves deprotonation of the malonate derivative (6) by piperidine (5) and attack by the formed carbanion (8) on the carbonyl subunit (9) as an aldol reaction that forms the product (10) of the addition step is illustrated in Fig. 2. After the proton transfer step between the protonated base (7) and compound (10), intermediate (11) forms and is then deprotonated to (12), which forms the elimination product (13) in the last step.

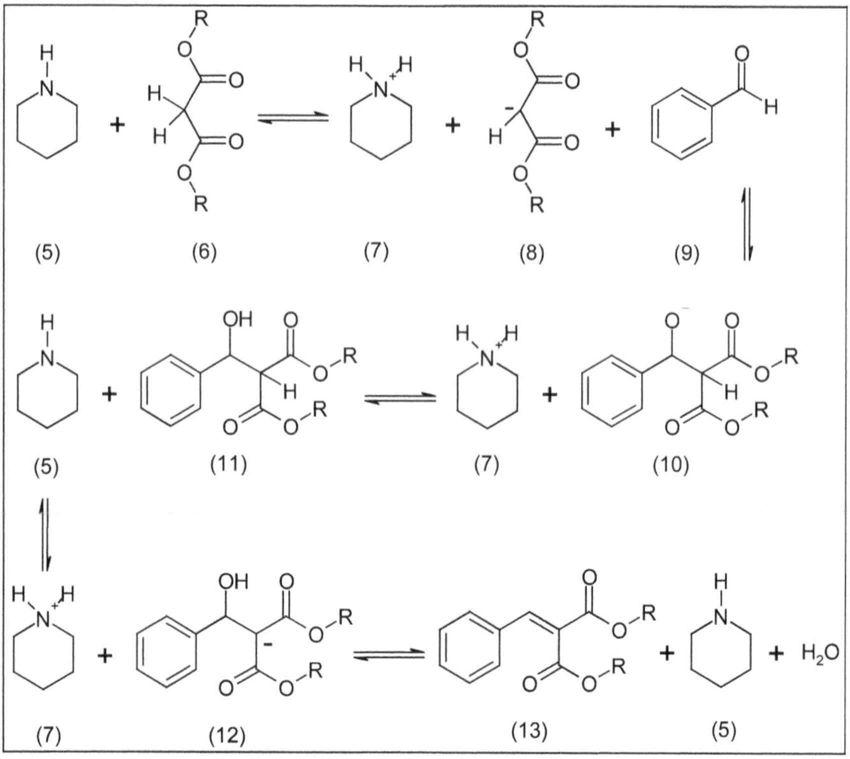

Fig. 2. General mechanism for the Knoevenagel reaction.

2. Green chemistry and new synthetic approaches

In the past two decades, classic organic chemistry had been rewritten around new approaches that search for products and processes in the chemical industry that are environmentally acceptable (Okkerse & Bekkum, 1999; Sheldon et al., 2007). With the emergence of Green Chemistry, a term coined in 1993 by Anastas at the US Environmental Protection Agency (EPA), a set of principles was proposed for the development of environmentally safer products and processes: waste prevention instead of remediation; atom efficiency; less hazardous/toxic chemicals; safer products by design; innocuous solvents and auxiliaries; energy efficiency by design; preference for renewable raw materials; shorter syntheses; catalytic rather than stoichiometric reagents; products designed for degradation; analytical methodologies for pollution prevention; and inherently safer processes (Anastas & Warner, 2000).

Consequently, many classic reactions, such as the Knoevenagel reaction, have been studied based upon the green chemistry perspective, which is very important in the context of the pharmaceutical industry. Currently, two indicators are used to evaluate environmental acceptability of products and chemical processes. The first is the Environmental factor (E factor), which measures the mass ratio of kg of waste to kg of desired product, as described by Sheldon in 1992 (Sheldon, 2007). The second indicator is a measure of atom economy based on the ratio of the molecular weight of the desired product to the sum of the molecular weights of all stoichiometric reagents. This indicator enables the evaluation of atom utilisation in a reaction (Trost, 1991). As illustrated in Table 1, the pharmaceutical industry produces 25->100 kg of waste per kg of drug produced, which is the worst E factor observed among the surveyed industrial sectors (Sheldon, 2007). This result is problematic as the pharmaceutical market is among the major sectors of the global economy, accounting for US $ 856 billion in 2010 (Gatyas, 2011a).

Industrial sector	Annual product tonnage	kg waste/ kg product
Oil refining	10^6-10^8	ca. 0.1
Bulk chemicals	10^4-10^6	<1-5
Fine chemicals	10^2-10^4	5->50
Pharmaceuticals	$10-10^3$	25->100

Table 1. The E Factor for selected industrial sectors, left justified.

Among the 20 top-selling drugs of 2010, atorvastatin (14) is at the top of the list, corresponding to US $ 12.6 billion in sales (Gatyas, 2011b). One step in the synthesis of atorvastatin (14) (Fig. 3) uses a Knoevenagel condensation between methylene compound (15) and benzaldehyde (9) to produce an intermediate (16) in yields of 85.0% (Li et al., 2004; Roth, 1993).

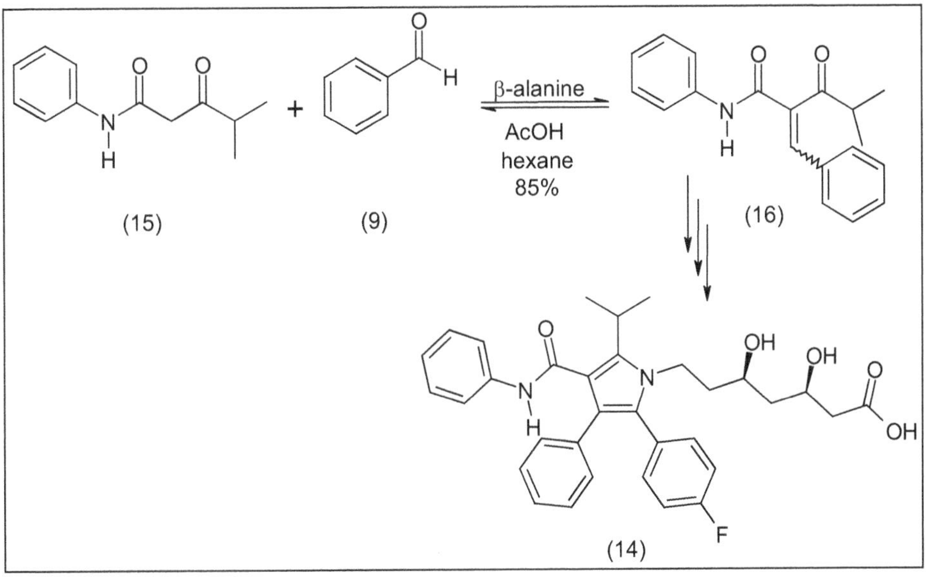

Fig. 3. A Knoevenagel condensation used during the synthesis of atorvastatin (14).

In addition to atorvastatin (14), many others drugs and pharmacological tools use the Knoevenagel reaction during their syntheses. Figure 4 illustrates the synthesis of pioglitazone (17), a benzylthiazolidinedione derivative approved as a drug for the management of diabetes (Madivada et al., 2009). In this synthesis, the key intermediate (20) was formed in yields of 94.5% through the piperidine-catalysed reaction of aldehyde intermediate (18) and 2,4-thiozolidinedione (19) (Madivada et al., 2009).

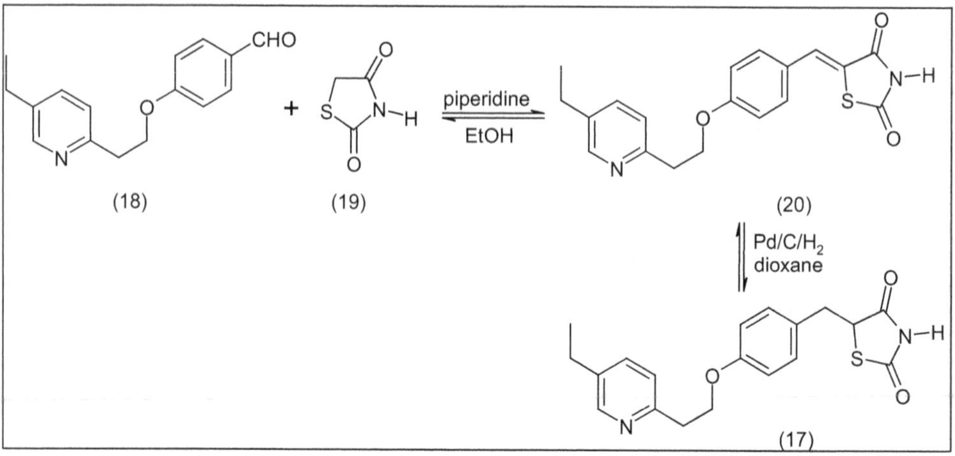

Fig. 4. Selected steps of the pioglitazone (17) synthesis.

AMG 837 (21) is a novel agonist of GPR40; this compound is being investigated as apotentially new therapeutic agent for the treatment of type 2 diabetes (Walker et al., 2011). As shown in Fig. 5, the synthetic route for AMG 837 (21) involves the production of intermediate (24), which is formed in yields of 97.0% via the reaction between aldehyde (22) with Meldrum's acid (23), using water/toluene (10/1) as a catalytic solvent (Walker et al., 2011).

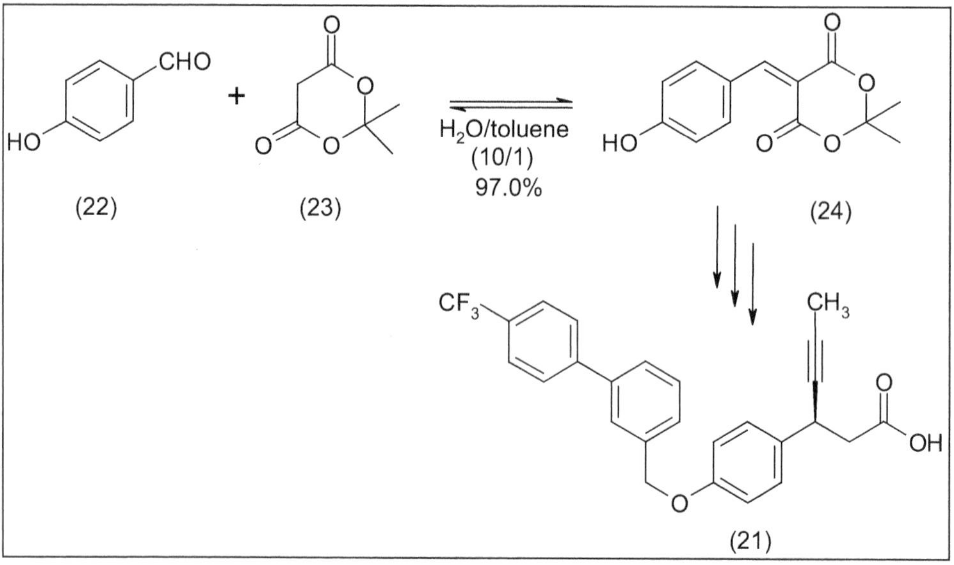

Fig. 5. AMG 837 (21) synthesis.

MDL 103371 (25) is an *N*-methyl-*D*-aspartate-type glycine receptor antagonist for the treatment of stroke (Watson et al., 2000). As illustrated in Fig. 6, synthesis of MDL 103371 (25) involves production of key intermediate (28) in yields of 91.0% via the condensation of 4,6-dichloro-3-formyl-1*H*-indole-2-carboxylate (26) with 3-nitrophenylacetonitrile (27); this piperidine-catalysed step is carried out using ethanol under reflux conditions for 70 hours (Walker et al., 2011).

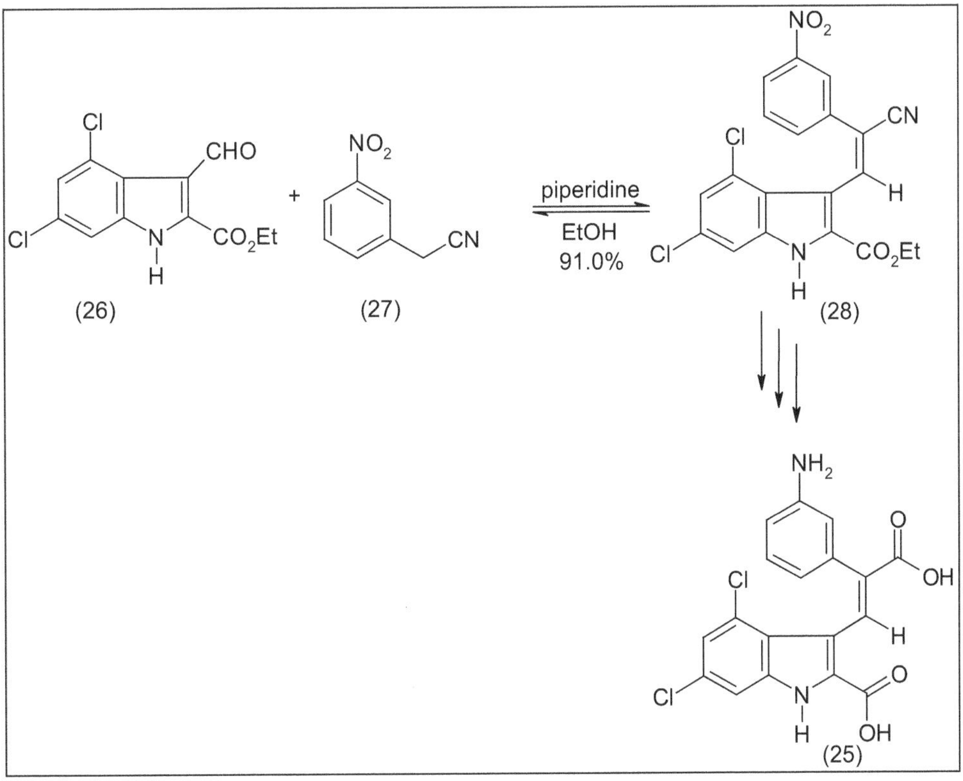

Fig. 6. MDL 103371 (25) synthesis.

(*S*)-(+)-3-Aminomethyl-5-methylhexanoic acid, or pregabalin (29), is a lipophilic GABA (γ-aminobutyric acid) analogue used for the treatment of several central nervous system (CNS) disorders, such as epilepsy, neuropathic pain, anxiety and social phobia (Martinez et al., 2008). As shown in Fig. 7, intermediate (32), which is produced during the synthesis of pregabalin (29), is formed in yields of 95.0% from the reaction between isovaleraldehyde (30) and diethyl malonate (31) using acetic acid as the solvent and di-*n*-propylamine as the catalyst.

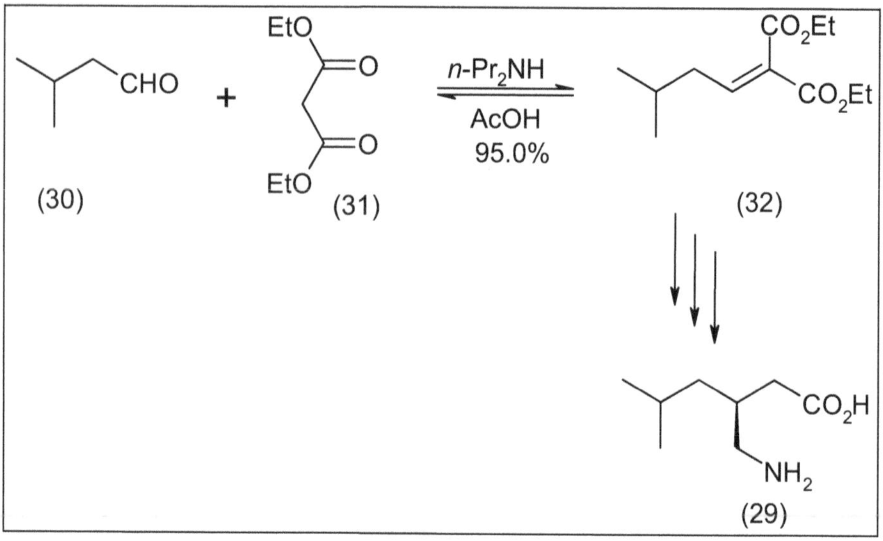

Fig. 7. Pregabalin (29) synthesis.

(E)-4-Cyclobutyl-2-[2-(3-nitrophenyl)ethenyl] thiazole, or Ro 24-5913 (33), is a leukotriene antagonist that has been utilised as a pharmacological tool to study asthma as well as other inflammatory diseases (Kuzemko et al., 2007). One method used to prepare Ro 24-5913 (33), illustrated in Fig. 8, is under Doebner conditions in which 3-nitrobenzaldehyde (34) reacts in the first step with malonic acid to produce intermediate (36) in yields of 65.6% (Kuzemko et al., 2007).

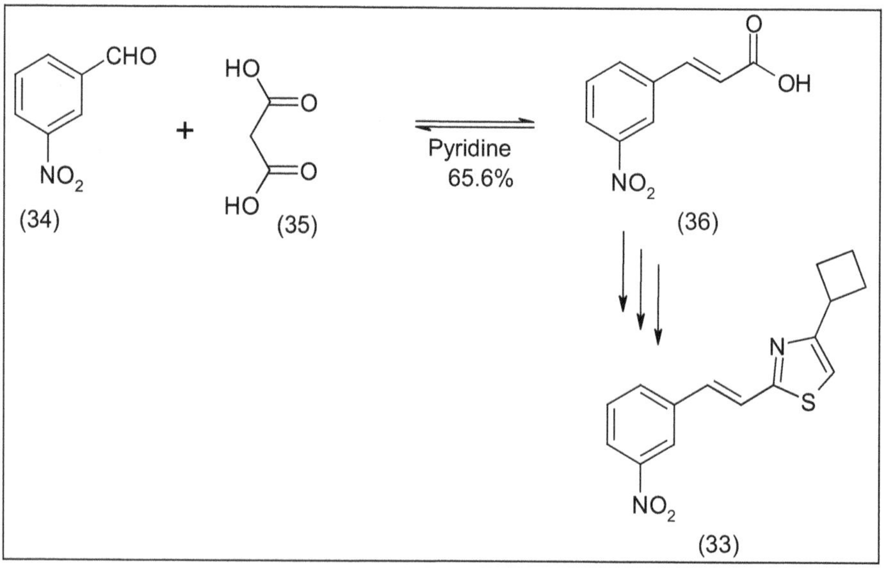

Fig. 8. Ro 24-5913 (33) synthesis.

Coartem is an antimalarial drug that is a combination of artemether and lumefantrine (37) (Beulter et al., 2007). This combination greatly benefits patients because it facilitates treatment compliance and supports optimal clinical effectiveness. As shown in Fig. 9, crude lumefantrine (37) was produced in yields of 88.0% via the reaction of 4-chlorobenzaldehyde (39) with methylene compound (38) in ethanol with sodium hydroxide as a catalyst. After crystallisation in heptane, pure lumefantrine (37) was generated in yields of 93.0% (Beulter et al., 2007).

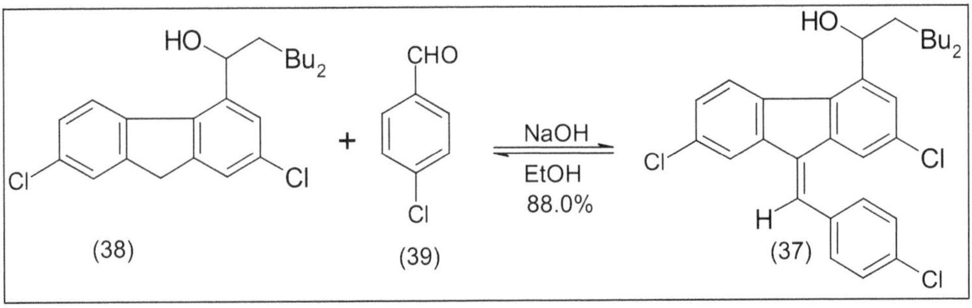

Fig. 9. Lumefantrine (37) synthesis.

Entacapone (40) is a catechol-O-methyltransferase (COMT) inhibitor used in combination with L-DOPA for the treatment of Parkinson's disease (Mukarram et al., 2007). This combination prevents L-DOPA degradation through COMT inhibition. As illustrated in Fig. 10, this drug (40) is synthesised in yields of 73.0% from aldehyde (41) and methylene compound (42) in ethanol with a piperidine catalyst (Mukarram et al., 2007).

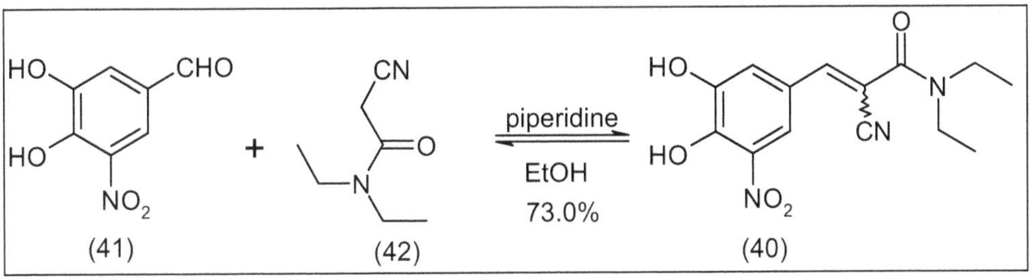

Fig. 10. Entacapone (40) synthesis.

The examples illustrated above for selected drug syntheses emphasise the on-going necessity of finding new approaches to carry out classic reactions that are essential to developing environmentally responsible products and chemical processes. Some new, green approaches are presented below.

2.1 Microwave-promoted Knoevenagel reactions

Microwave irradiation is a method used to speed up reactions with potential uses under the guidelines of Green Chemistry principles. Microwave radiation utilises wavelengths of 0.001 – 1 m and frequencies of 0.3 – 300 GHz. When a polar organic reaction is irradiated in a microwave, energy is transferred to the sample, and the result is an increase in the rate of reaction. The transference of energy from microwave radiation to the sample is accomplished through dipolar polarisation and conduction mechanisms (Lidström et al. 2001; Loupy, 2002). As illustrated in Fig. 11, the coumarinic derivative (42) is produced in yields of 75.0% after eight minutes of irradiation. This reaction was carried out using aldehyde (43) and methylene compound (44) and was catalysed by piperidine without solvent present (Bogdal, 1998). Following the Knoevenagel condensation, the transesterification reaction to form the ring quickly occurs.

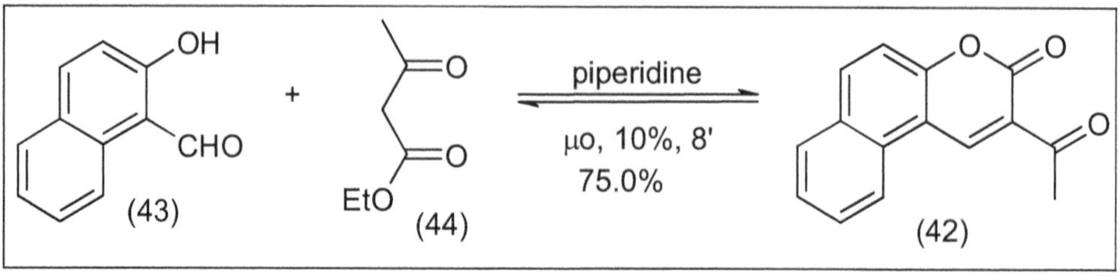

Fig. 11. Coumarinic derivative (42) synthesis.

There are many examples in the literature involving the use of microwave radiation to promote Knoevenagel reactions. In these examples, several different aldehydes, methylene compounds and catalysts were used for the syntheses involving cinnamic acids on silica gel (Kumar et al., 2000), ammonium acetate (Kumar et al., 1998; Mitra et al., 1999) and lithium chloride as catalysts (Mogilaiah & Reddy, 2004).

2.2 Clays as catalysts for Knoevenagel reactions

Clays are abundant in nature, and their high surface area, utility as supports and ion-exchange properties have been exploited for catalytic applications (Dasgupta & Török, 2008; Varma, 2002). As shown in Fig. 12, the product of Knoevenagel reaction (45) from the reaction between ninhydrin (46) and malononitrile (47) can be formed in yields of 85.0% after five minutes. This reaction was carried out at room temperature without solvent using K10 as a catalyst (Chakrabarty et al., 2009).

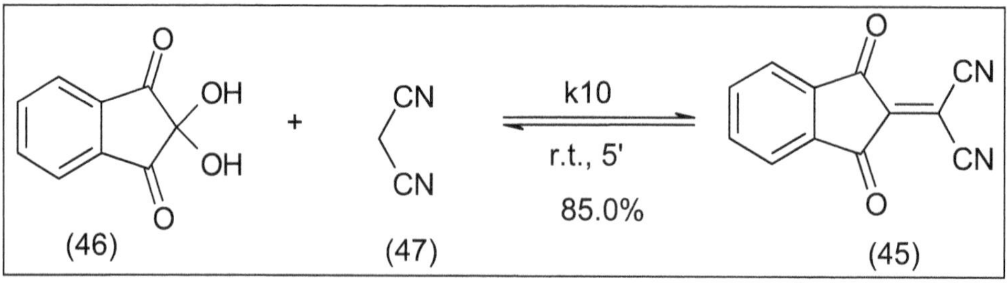

Fig. 12. Knoevenagel product (45) synthesis.

Other Knoevenagel reactions between aromatic aldehydes and malononitrile (47) have also performed successfully without solvent using calcite or fluorite catalysts prepared using a ball mill (Wada & Suzuki, 2003).

2.3 The use of ionic liquids in Knoevenagel reactions

In recent years, ionic liquids (ILs) have attracted increasing interest as environmentally benign solvents and catalysts due to their relatively low viscosities, low vapour pressures and high thermal and chemical stabilities (Hajipour & Rafiee, 2010; Wasserscheid & Welton, 2002). ILs have been successfully used in a variety of reactions.

As illustrated in Fig 13, the pyrazolonic compound (48) was produced in yields of 71.0% from the reaction between benzaldehyde (9) and 3-methyl-1-phenylpyrazolin-5-(4H)-one (49) after 30 minutes using ethylammonium nitrate as an ionic liquid at room temperature (Hangarge et al., 2002).

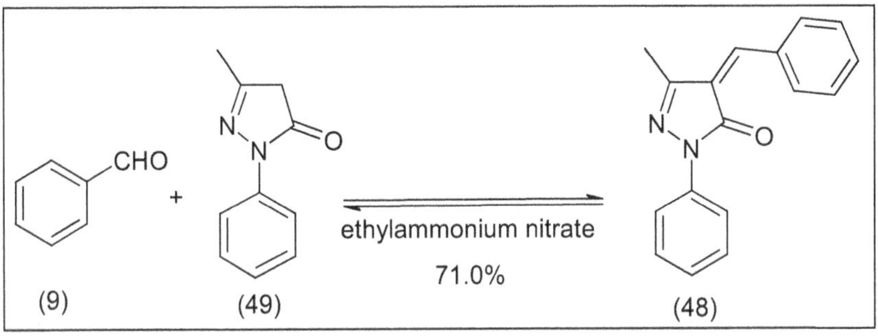

Fig. 13. Compound (48) synthesis.

Other reactions between aromatic aldehydes and methylene compounds that were catalysed by 1,3-dimethylimidazolium methyl sulphate [MMIm][MSO4] and 2.16% water have been carried out in good yields (Verdía et al., 2011),

2.4 Catalysis of Knoevenagel reactions using biotechnology

Historically, microorganisms have been of enormous social and economic importance (Liese et al., 2006). In the pharmaceutical industry, companies are using biotechnology to develop 901 medicines and vaccines targeting more than 100 diseases (Castellani, 2001a). In 2010, 26 new treatments were approved, and five of these treatments were based on biotechnology (Castellani, 2001b).

Using a biotechnology-based approach, coumarin (50) was produced in yields of 58.0% when the reaction was catalysed by alkaline protease from *Bacillus licheniformis* (BLAP) in a DMSO:H_2O (9:1) solvent at a temperature of 55°C (Fig. 14) (Wang et al., 2011).

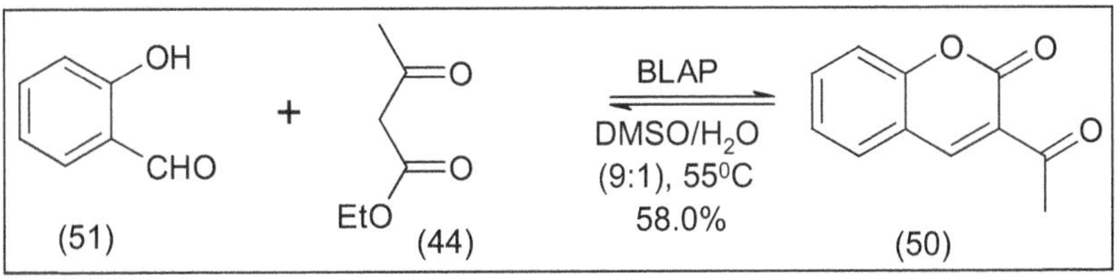

Fig. 14. Coumarin (50) synthesis using BLAP.

Because cells are chemical systems that must conform to all chemical and physical laws, whole microorganisms may be used (Alberts et al., 2002). Figure 15 illustrates examples of other Knoevenagel products (52) and (53) resulting from reactions between benzaldehyde (9) and methylene compounds (44) and (19) that were catalysed by baker's yeast. These reactions were carried out under mild conditions, e.g., room temperature and in ethanol as the solvent, with moderate to good yields (Pratap et al., 2011).

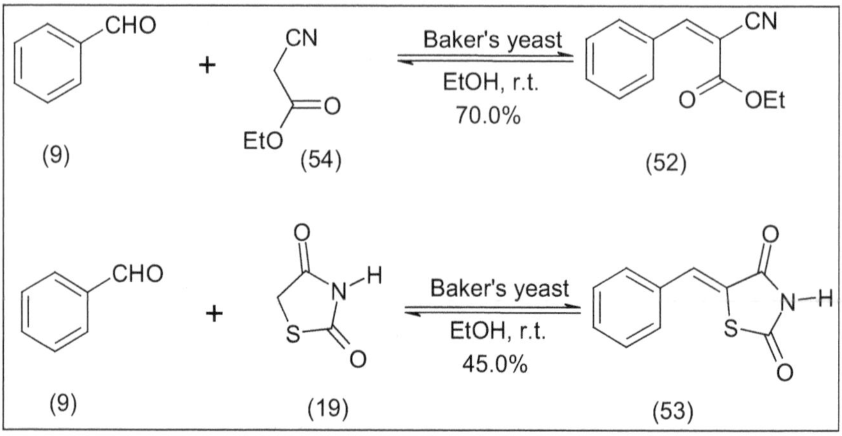

Fig. 15. Knoevenagel reactions using a biotechnology-based approach.

2.5 Knoevenagel reactions in water

Water as a solvent is not only inexpensive and environmentally benign but also provides completely different reactivity (Li & Chen, 2006). It has been suggested that the effect of water on organic reactions may be due to the high internal pressure exerted by a water solution, which results from the high cohesive energy of water (Breslow, 1991).

As illustrated in Fig. 16, the Knoevenagel reaction product (55) is formed in yields of 97.0% when condensation between aldehyde (56) and malononitrile (47) was carried out in water at a temperature of $65^{0}C$ in the absence of catalyst (Bigi et al., 2000).

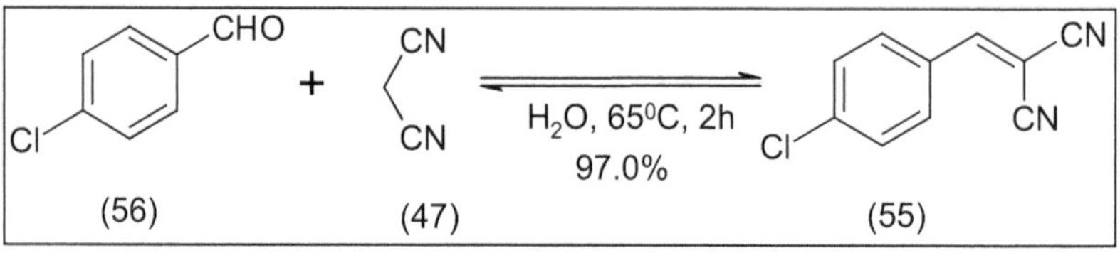

Fig. 16. A Knoevenagel reaction carried out in water.

The product of the reaction between vanillin (58) and ethyl cyanoacetate (54) was formed in yields of 84.5% in water at room temperature (Fig. 17) (Gomes et al., 2011). This compound was patented in 2007 by Merck & Co. for use in sunscreen compositions containing a UVA sunscreen, photostabliser and antioxidant. The reaction was carried out using piperidine as the catalyst and acetic acid/benzene as the solvent under reflux conditions for approximately 90 minutes, producing yields of 95.0% (Ratan, 2007). Thus, it's clear that there are green approaches for carrying out organic reactions in water to prepare compounds of industrial interest.

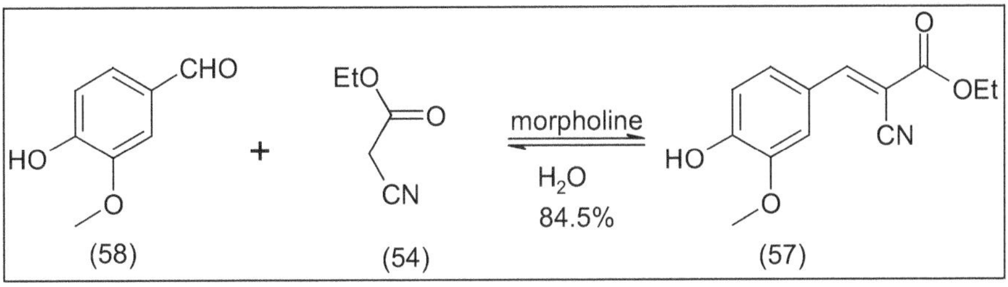

Fig. 17. Morphonile-catalysed synthesis of sunscreen in water.

Entacapone (40), a COMT inhibitor drug whose synthesis is illustrated above in Fig. 10, is another example of the synthesis of important industrial compounds using green conditions. As shown in Fig. 18, Knoevenagel reaction product (59) is formed in yields of 88.0% after two hours under reflux in water with piperidine as a catalyst (McCluskey, 2002).

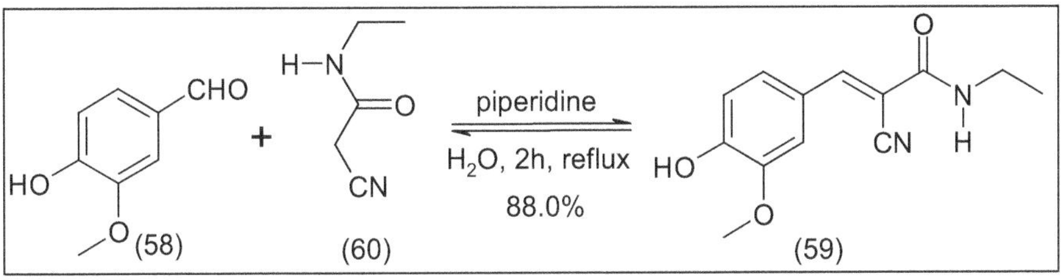

Fig. 18. Compound (59) synthesis.

There are others examples of Knoeveganel reactions carried out in water that are catalysed by L-histidine and L-arginine (Rhamati & Vakili, 2010). Isatin compounds (61) can also be produced in water at room temperature after fifteen minutes in yields of 75.0% (Fig. 19) (Demchuk, 2011).

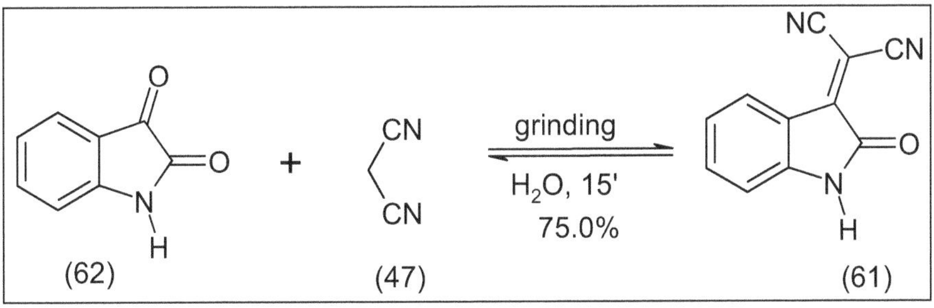

Fig. 19. Isatin compound (61) synthesis.

Knoevenagel reactions can also be used to assemble a benzo[b]pyrane [4,3-d][1,2]oxazine-2-oxide skeleton (63) and (64) via a domino-effect Knoevenagel–Diels–Alder process (Fig. 20)

(Amantini, 2001). When the prenylated phenolic aldehyde (65) reacts with methylene compound (66) in water at room temperature for three hours, Knoevenagel intermediate (67) forms, which then reacts to form Diels-Alder product (63) and (64) in yields of 75.0% at a 16:1 ratio (Amantini, 2001).

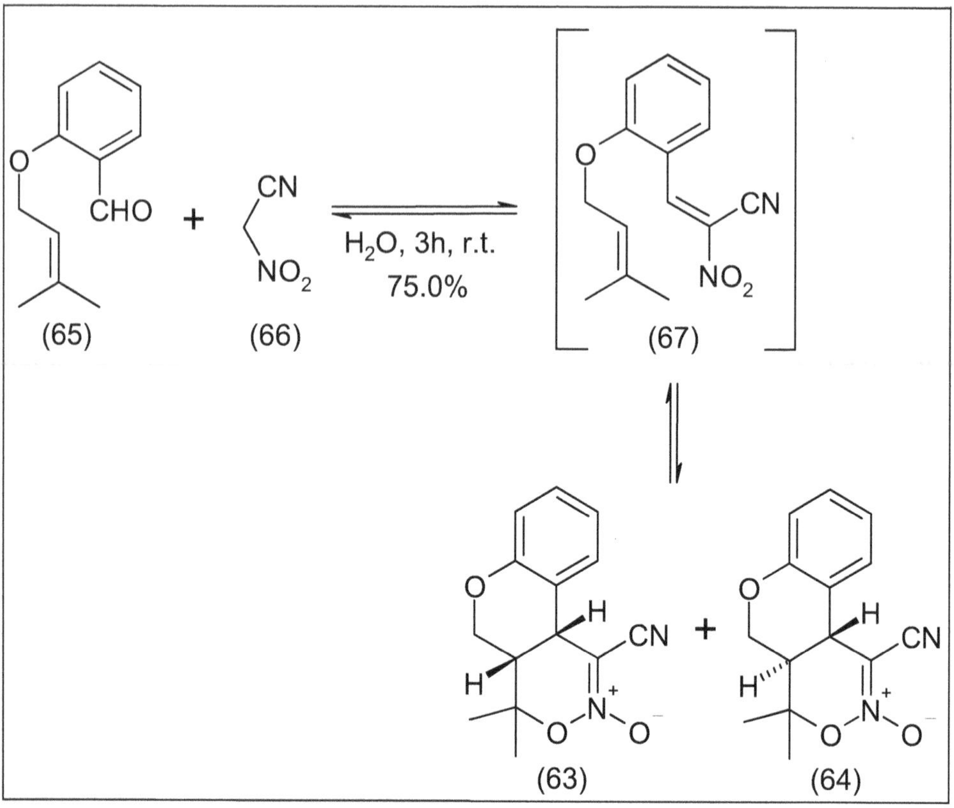

Fig. 20. Synthesis of benzo[b]pyrane [4,3-d][1,2]oxazine-2-oxide skeletons (63) and (64).

2.6 Ultrasound-catalysed Knoevenagel reactions

The application of ultrasound waves triggers high-energy chemistry, which is thought to occur through the process of acoustic cavitation, i.e., the formation, growth and implosive collapse of bubbles in a liquid. During cavitational collapse, intense heating of the bubbles occurs (Suslick, 1990).

The piperidine-catalysed reaction between piperonal (69) and malonic acid (35) at room temperature with pyridine as the solvent was carried out under ultrasound irradiation, and Knoevenagel reaction product (68) formed in yields of 91.0% after three hours (Fig. 21) (McNulty et al., 1998). When carried out under reflux conditions, the same reaction forms the Knoevenagel reaction product in yields of 52.0% after three hours (McNulty et al., 1998).

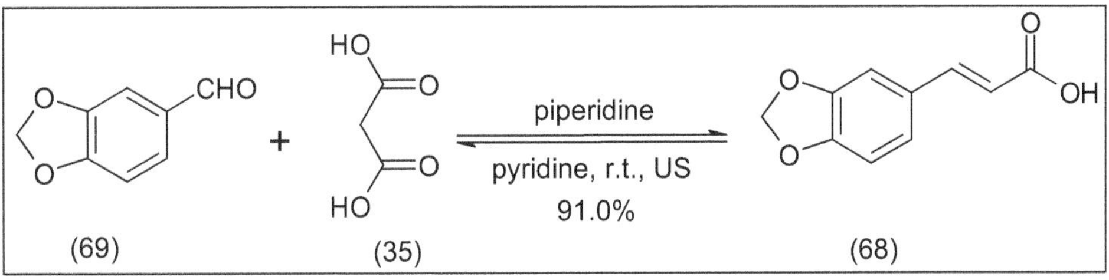
HO
CHO
O
O
+
O
O
HO
piperidine
pyridine, r.t., US
91.0%
O
O
OH
O
(69)
(35)
(68)

Fig. 21. Ultrasound-catalysed synthesis of Knoevenagel reaction product (68).

Figure 22 illustrates reactions between benzaldehyde (9) and coumarin (71) that can also be conducted in water under ultrasound irradiation at a temperature of 40ºC for 90 minutes, forming product (70) in yields of 88.0% (Method B) (Palmisano et al., 2011). In the absence of ultrasound irradiation, formation of product (70) occurs in yields of 62.0% under anhydric conditions (Method A) (Palmisano et al., 2011).

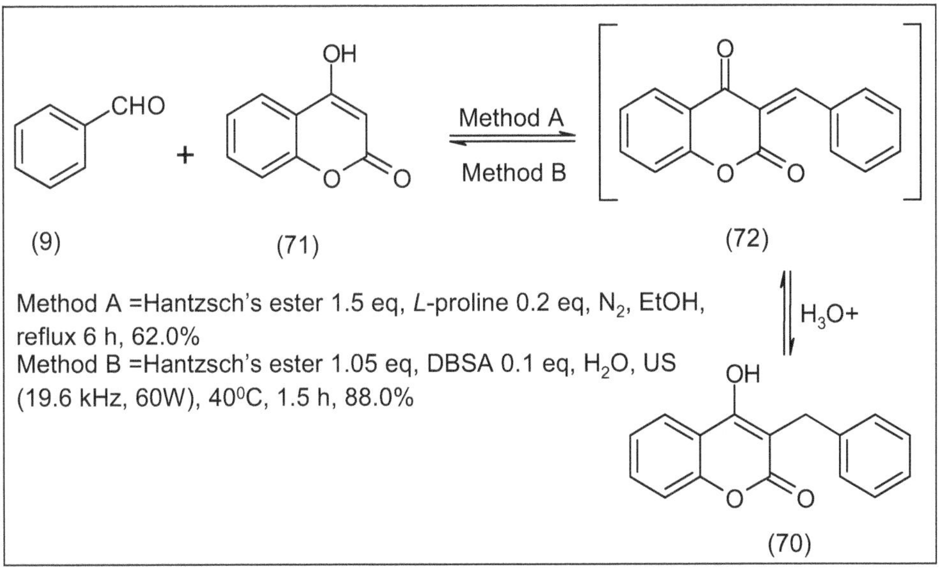
OH
CHO
+
O
O
Method A
Method B
O
O
O
(9)
(71)
(72)
Method A =Hantzsch's ester 1.5 eq, L-proline 0.2 eq, N2, EtOH, reflux 6 h, 62.0%
Method B =Hantzsch's ester 1.05 eq, DBSA 0.1 eq, H2O, US (19.6 kHz, 60W), 40ºC, 1.5 h, 88.0%
H3O+
OH
O
O
(70)

Fig. 22. Synthesis of coumarinic compound (70) with and without ultrasound irradiation.

2.7 Solvent-free Knoevenagel reactions

As mentioned previously, the reduction or elimination of volatile organic solvents in organic syntheses is one of the main goals in green chemistry. Solvent-free organic reactions result in syntheses that are simpler and less energy-intensive, and these conditions also reduce or eliminate solvent waste, hazards, and toxicity (Tanaka, 2003).

One method of solvent-free organic synthesis uses high pressure, as shown in Fig. 22 (Jenner, 2001). The piperidine-catalysed reaction between 2-butanone (75) and ethyl cyanoacetate (54) was carried out using two methods in which formation of E(73)/Z(74) Knoevenagel reaction products was observed. When the pressure was increased from 0.10

MPa to 300 MPa, the yield increased from 28.1% to 99.0%. However, significant changes were not observed in the ratio of E(73)/Z(74) Knoevenagel reaction products (Jenner, 2001).

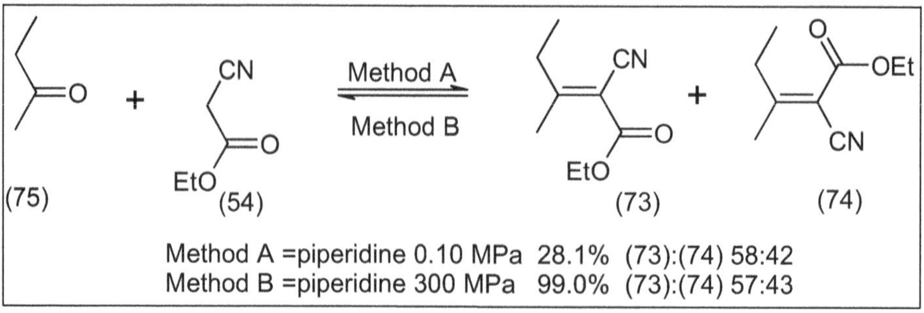

Method A =piperidine 0.10 MPa 28.1% (73):(74) 58:42
Method B =piperidine 300 MPa 99.0% (73):(74) 57:43

Fig. 23. High-pressure Knoevenagel reactions.

Solvent-free Knoevenagel reactions have also been carried out using a mortar and pestle. Under these conditions, the reaction between benzaldehyde (9) diethyl malonate (31), which was catalysed by triethylbenzylammonium chloride (TEBA), resulted in product yields of 87.5% after ten minutes (Rong et al., 2006). Similarly, the domino-effect Friedlander condensation reaction, which can also be conducted using a mortar and pestle, was observed between aldehyde (77) and methylene compound (78). This sodium fluoride-catalysed reaction formed the aromatic Knoevenagel reaction product (76) in yields of 92.0% after eight minutes (Fig. 24) (Mogilaiah & Reddy, 2003).

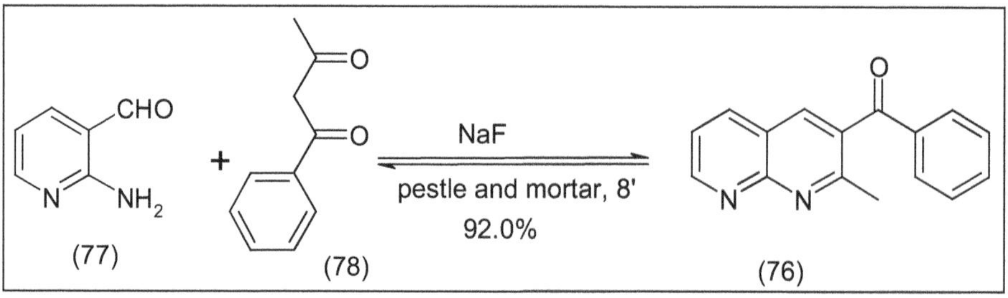

Fig. 24. Synthesis of the Friedlander condensation product (76).

2.8 Knoevenagel reactions using solid phase organic synthesis

An innovative and important field of organic synthesis involves the use of solid phase organic synthesis (Czarnik, 2001). This new methodology was introduced by Merrifield in 1963 when he used it to synthesise amino acids (Merrifield, 1963). Solid phase organic synthesis uses insoluble polymers that covalently bond organic substrates to the solid surface until the synthesis is complete, at which point the compound of interest is separated from the solid matrix (Czarnik, 2001).

This approach has been used to synthesise coumaric compound (79) from the reaction of aldehyde (81) and methylene compound (80) bonded to Wang resin. The reaction was complete after sixteen hours under Doebner conditions, as illustrated in Fig. 25 (Xia et al., 1999).

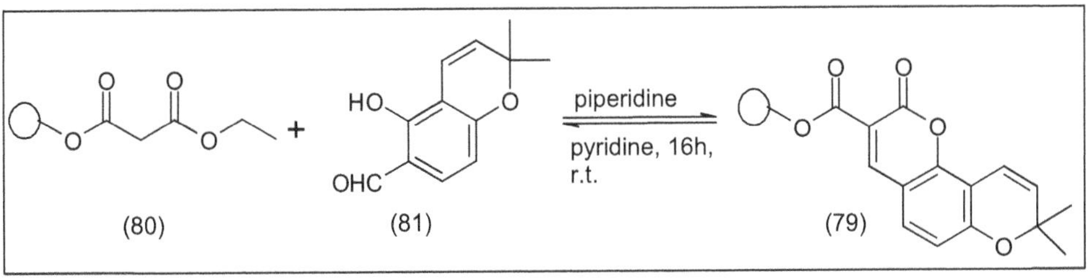

Fig. 25. Synthesis of coumarin (79) via solid phase organic synthesis.

Solid phase Knoevenagel reactions were also utilised to produce triphostin protein tyrosine kinases inhibitors. As illustrated in Fig. 26, the piperidine-catalysed reaction between 4-hydroxybenzaldehyde (22) and a resin-bonded methylene compound (83) was carried out using DMF:MeOH (10:1) as the solvent over a period of twelve hours (Guo et al., 1999).

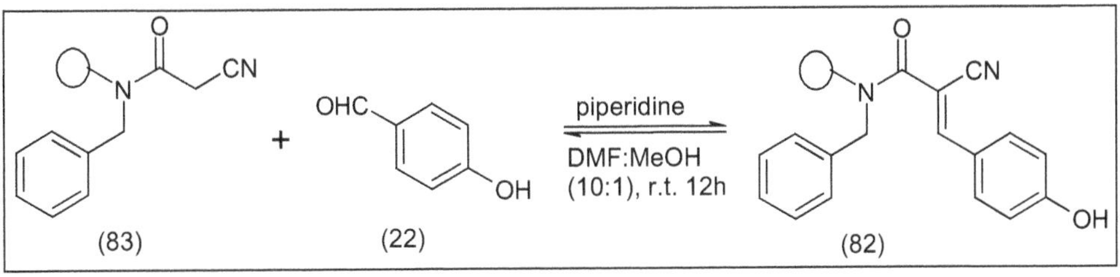

Fig. 26. Synthesis of triphostin (82) via solid phase organic synthesis.

3. Conclusions

As illustrated by the examples presented herein, classic reactions such as the Knoevenagel condensation can be modernised through new approaches related to Green Chemistry. Particularly in the area of drug synthesis, these new approaches have been being very useful in the development of more environmentally supportable products and chemical processes in the pharmaceutical industry, which works with compounds with high added values.

4. Acknowledgments

The author is grateful to FAPEG, INCT-INOFAR and UFG for financial support.

5. References

Alberts, B., Johonson, A., Lewis, J., Raff, M., Roberts, K. & Walter, P. (2002). *Molecular Biology of The Cell*, Garland Science, ISBN 0-8153-3218-1, New York, USA

Amantini, D., Fringuelli, F., Piermatti, O., Pizzo, F. & Vaccaro, L. (2001). Water, a clean, inexpensive, and re-usable reaction medium. One-pot synthesis of (E)-2-aryl-1-cyano-1-nitroethenes. *Green Chemistry*, Vol.3, No.5, (September 2001), pp. 229-232, ISSN 1463-9262

Anastas, P. T. & Warner, J.C. (2000). *Green Chemistry : Theory and Practice*, Oxford University Press, ISBN 0-19-850698-8, New York, USA

Beutler, U., Fuenfschilling, P. C. & Steinkemper, A. (2007). An Improved Manufacturing Process for the Antimalaria Drug Coartem. Part II. *Organic Process Research & Development*, Vol.11, No.3, (June 2007), pp. 470-476, ISSN 341-345

Bigi, F., Conforti, M. L., Maggi, R., Piccinno, A. & Sartori, G. (2000). Clean synthesis in water: uncatalysed preparation of ylidenemalononitriles. *Green Chemistry*, Vol.2, No.3, (May 2000), pp. 101-103, ISSN 1463-9262

Bogdal, D. (1998). Coumarins: Fast Synthesis by Knoevenagel Condensation under Microwave Irradiation. *Journal of Chemical Research (S)*, (March 1998), pp. 468-469, ISSN 1747-5198

Breslow, R. (1991). Hydrophobic Effects on Simple Organic Reactions in Water. *Accounts of Chemical Research*, Vol.24, No.6, (June 1991), pp. 159-164, ISSN 0001-4842

Burgess, K. (2000). *Solid-Phase Organic Synthesis*, John Wiley & Sons, Inc., ISBN 0-471-31825-6, New York, USA

Castellani, J. J. (September 2011a). Medicines in development Biotechnology, 23.09.2011, Available from
http://www.phrma.org/sites/default/files/1776/biotech2011.pdf

Castellani, J. J. (September 2011b). New Drug Approvals in 2010, 23.09.2011, Available from
http://www.phrma.org/sites/default/files/422/newdrugapprovalsin2010.pdf

Chakrabarty, M., Mukherji, A., Arima, S., Harigaya, Y. & Pilet, G. (2009). Expeditious reaction of ninhydrin with active methylene compounds on montmorillonite K10 clay. *Monatshefte für Chemie*, Vol.140, No.2, (February 2009), pp. 189-197, ISSN 0026-9247

Czarnik, A. W. (2001). *Solid-Phase Organic Syntheses*, John Wiley & Sons, Inc., ISBN 0-471-31484-6, New York, USA

Dasgupta, S. & Török, B. (2008). Application of Clay Catalysts in Organic Synthesis. A Review. *Organic Preparations and Procedures International*, Vol.40, No.1, (February 2008), pp. 1-65, ISSN 0030-4948

Demchuk, D. V., Elinson, M. N. & Nikishin, G. I. (2011). 'On water' Knoevenagel condensation of isatins with malononitrile. *Mendeleev Communications*, Vol.21, No.4, (July-August 2011), pp. 224-225, ISSN 0959-9436

Doebner, O. (1900). Synthese der Sorbinsäure. *Berichte der Deutschen Chemischen Gesellschaft*, Vol.33, No.2, (Mai-August 1900), pp. 2140-2142, ISSN 0365-9631

Gatyas, G. (August 2011a). Global Pharmaceutical Sales 2003-2010, 10.08.2011, Available from
http://www.imshealth.com/deployedfiles/ims/Global/Content/Corporate/Press%20Room/Top-line%20Market%20Data/2010%20Top-line%20Market%20Data/Total_Market_2003-2010.pdf

Gatyas, G. (August 2011b). Top 20 Global Products, 2010, 10.08.2011, Available from
http://http://www.imshealth.com/deployedfiles/ims/Global/Content/Corporate/Press%20Room/Top-line%20Market%20Data/2010%20Top-line%20Market%20Data/Top_20_Global_Products.pdf

Gomes, M. N., de Oliveira, C. M. A., Garrote, C. F. D., de Oliveira, V. & Menegatti, R. (2011). Condensation of Ethyl Cyanoacetate with Aromatic Aldehydes in Water,Catalyzed by Morpholine. *Synthetic Communication,* Vol.41, No.1, (January 2011), pp. 52-57, ISSN 0039-7911

Guo, G., Arvanitis, E. A., Pottorf, R. S. & Player, M. P. (2003). Solid-Phase Synthesis of a Tyrphostin Ether Library. *Journal of Combinatorial Chemistry,* Vol.5, No.4, (January 2011), pp. 408-413, ISSN 1520-4766

Hajipour, A. R. & Rafiee, F. (2010). Acidic Bronsted Ionic Liquids. *Organic Preparations and Procedures International,* Vol.42, No.4, (July 2010), pp. 285-362, ISSN 0030-4948

Hangarge, R. V., Jarikote, D. V. & Shingare, M. S. (2002). Knoevenagel condensation reactions in an ionic liquid. *Green Chemistry,* Vol.4, No.3, (May 2002), pp. 266-268, ISSN 1463-9262

Jenner, G. (2001). Steric effects in high pressure Knoevenagel reactions. *Tetrahedron Letters,* Vol.42, No.2, (January 2001), pp. 243-245, ISSN 0040-4039

Knoevenagel, E. (1894). Ueber eine Darstellungsweise der Glutarsäure. *Berichte der Deutschen Chemischen Gesellschaft,* Vol.27, No.2, (Mai-August 1894), pp. 2345-2346, ISSN 0365-9631

Kumar, H. M. S., Reddy, B. V. S., Reddy, P. T., Srinivas, D. & Yadav, J. S. (2000). Silica Gel Catalyzed Preparation of Cinnamic Acids Under Microwave Irradiation. *Organic Preparations and Procedures Int.,* Vol.32, No.1, (February 2000), pp. 81-83, ISSN 0030-4948

Kumar, H. M. S., Subbareddy, B. V., Anjaneyulu, S. & Yadav, J. S. (1998). Non Solvent Reaction: Ammonium Acetate Catalyzed Highly Convenient Preparation of Trans-Cinnamic Acids. *Synthetic Communication,* Vol.28, No.20, (August 1998), pp. 3811-3815, ISSN 0039-7911

Kuzemko, M. A., Van Arnum, S. D. & Niemczyk, H. J. (2007). A Green Chemistry Comparative Analysis of the Syntheses of (E)-4-Cyclobutyl-2-[2-(3-nitrophenyl)ethenyl] Thiazole, Ro 24-5904. *Organic Process Research & Development,* Vol.11, No.3, (June 2008), pp. 470-476, ISSN 1083-6160

Laue, T. & Plagens, A. (2005). *Named Organic Chemistry,* John Wiley & Sons Ltd, ISBN 0-470-01040-1, Wolfsburg, Germany

Li, C. & Chen, L. (2006). Organic Chemistry in Water. *Chemical Society Reviews,* Vol.35, No.1, (January 2006), pp. 68-82, ISSN 0306-0012

Liese, A., Seelbach, K. & Wandrey, C. (2006). *Industrial Biotransformations,* WILEY-VCH Verlag GmbH & Co. KGaA, ISBN 3-527-31001-0, WILEY-VCH Verlag GmbH & Co. KGaA, Weinheim, Germany

Li, J. J., Johnson, D. S., Sliskovic, D. R. & Roth, B. D. (2004). *Contemporary Drug Synthesis,* Wiley- Interscience, ISBN 0-471-21480-9, New Jersey, USA

Lidström, P., Tierney, J., Wathey, B. & Westman, J. (2001). Microwave assisted organic synthesis – a review. *Tetrahedron,* Vol.57, No.45, (November 2001), pp. 9225-9283, ISSN 0040-4020

Loupy, A. (2002). *Microwaves in Organic Synthesis,* WILEY-VCH Verlag GmbH & Co. KGaA, ISBN 3-527-30514-9, Weinheim, Germany

Madivada, L. R., Anumala, R. R, Gilla, G., Alla, S., Charagondla, K., Kagga, M., Bhattacharya, A. & Bandichhor, R. (2009). An Improved Process for Pioglitazone and Its Pharmaceutically Acceptable Salt. *Organic Process Research & Development,* Vol.13, No.6, (December 2009), pp. 1190-1194, ISSN 1083-6160

Martinez, C. A., Hu, S., Dumond, Y., Tao, J., Kelleher, P. & Tully, L. (2008). Development of a Chemoenzymatic Manufacturing Process for Pregabalin. *Organic Process Research & Development,* Vol.12, No.3, (June 2008), pp. 392-398, ISSN 1083-6160

McCluskey, A., Robinson, P. J., Hill, T., Scottc, J. L. & Edwards, J. K. (2002). Green chemistry approaches to the Knoevenagel condensation: comparison of ethanol, water and solvent free (dry grind) approaches. *Tetrahedron Letters,* Vol.43, No.17, (April 2002), pp. 3117-3120, ISSN 0040-4039

McNulty, J., Steere, I. J. A. & Wolf, S. (1998). The Ultrasound Promoted Knoevenagel Condensation of Aromatic Aldehydes. *Tetrahedron Letters,* Vol.39, No.29, (October 2000), pp. 8013-8016, ISSN 0040-4039

Merrifield, R. K. (1963). Solid Phase Peptide Synthesis. I. The Synthesis of a Tetrapeptide. *Journal of the American Chemical Society,* Vol.85, No.14, (July 1963), pp. 2149-2154, ISSN 0002-7863

Mogilaiah, K. & Reddy, C. S. (2003). An Efficient Friedlander Condensation Using Sodium Fluoride as Catalyst in the Solid State. *Synthetic Communication,* Vol.33, No.18, (August 2003), pp. 3131-3134, ISSN 0039-7911

Mogilaiah, K. & Reddy, G. R. (2004). Microwave-Assisted Solvent-Free Synthesis of trans-Cinnamic Acids Using Lithium Chloride as Catalyst. *Synthetic Communication,* Vol.34, No.2, (January 2004), pp. 205-210, ISSN 0039-7911

Mukarram, S. M. J., Khan, R. A. R. & Yadav, R. P. (2007). Efficient Method for the Manufacture of (E)-Entacapone Polymorphic for a. United States Patent, 2007, Patent No: 0004935

Okkerse, C. and van Bekkum, V. (1999). From Fossil to Green. *Green Chemistry,* Vol.1, No.2, (April 1999), pp. 107-114, ISSN 1463-9262

Palmisano, G., Tibiletti, F., Penoni, A., Colombo, F., Tollari, S., Garella, D., Tagliapietra, S. & Cravotto, G. (2011). Ultrasound-enhanced one-pot synthesis of 3-(Het)arylmethyl-4-hydroxycoumarins in water. *Ultrasonics Sonochemistry,* Vol.18, No.2, (March 2011), pp. 652-660, ISSN 1350-4177

Pratap, U. R., Jawale, D. V., Waghmare, R. A., Lingampalle, D. L. & Mane, R. A. (2011). Synthesis of 5-arylidene-2,4-thiazolidinediones by Knoevenagel condensation catalyzed by baker's yeast. *New Journal of Chemistry,* Vol.35, No.1, (November 2011), pp. 49-51, ISSN 1144-0546

Rahmati, A. & Vakili, K. (2010). *L*-Histidine and *L*-arginine promote Knoevenagel reaction in water. *Amino Acids,* Vol.39, No.3, (June 2010), pp. 911-916, ISSN 0939-4451

Ratan, C. (2007). Sunscreen compositions containing a UV-A sunscreen and photostablizers and antioxidants. United States Patent, 2007, Patent No: 0059258

Ronga, L., Li, X., Wang, H., Shia, D., Tua, S., Zhuang, Q. (2006). Efficient Green Procedure for the Knoevenagel Condensation under Solvent-Free Conditions. *Synthetic Communication,* Vol.36, No.16, (September 2006), pp. 2407-2412, ISSN 0039-7911

Roth, B. D. (1993). [R-(R*R*)]-2-(4-fluorphenyl)-beta, delta –dihydroxy-5-(1-methylethyl-3-phenyl-4[(phenylamino) carbonyl]-1H-pyrrole-1-heptanoic acid, its lactone form and salts trereof. United States Patent, 1993, Patent No: 5273995

Sheldon, R. A. (2007). The E Factor: fifteen years on. Green Chemistry, Vol.9, No.12, (December 2007), pp. 1273-1283, ISSN 1463-9262

Sheldon, R. A., Arends, I. & Hanefeld, U. (2007). Green Chemistry and Catalysis, Wiley-VCH Verlag GmbH & Co. kGaA, ISBN 978-3-527-30715-9, Weinheim, Germany

Smith, B. & March, J. (2001). March's Advanced Organic Chemistry, John Wiley & Sons Ltd, ISBN 0-471-58589-0, New York, USA

Suslick, K. S. (1990). Sonochemistry, Science, Vol.247, No.4949, (March 1990), pp. 1439-1445, ISSN 0036-8075

Tanaka, K. (2003). Solvente-free Organic Synthesis, WILEY-VCH Verlag GmbH & Co. KGaA, ISBN 3-527-30612-9, Weinheim, Germany

Trost, B. M. (1991). The atom economy – a search for synthetic efficiency. Science, Vol.254, No.5037, (Dec 1991), pp. 1471-1477, ISSN 0036-8075

Varma, R. S. (2002). Clay and clay-supported reagents in organic synthesis. Tetrahedron, Vol.58, No.7, (February 2002), pp. 1235-1255, ISSN 0040-4020

Verdía, P., Santamarta, F. & Tojo, E. (2011). Knoevenagel Reaction in [MMIm][MSO4]: Synthesis of Coumarins. Molecules, Vol.16, No.6, (June 2011), pp. 4379-4388, ISSN 1320-3049

Wada, S. & Suzuki, H. (2003). Calcite and fluorite as catalyst for the Knoevenagel condensation of malononitrile and methyl cyanoacetate under solvent-free conditions. Tetrahedron Letters, Vol.44, No.2, (January 2003), pp. 399-401, ISSN 0040-4039

Walker, S. D., Borths, C. J., DiVirgilio, E., Huang, L., Liu, P., Morrison, H., Sugi, K., Tanaka, M., Woo, J. C. S & Faul, M. M. (2011). Development of a Scalable Synthesis of a GPR40 Receptor Agonist. Organic Process Research & Development, Vol.15, No.3, (March 2011), pp. 570-580, ISSN 1083-6160

Wang, C., Guan, Z. & He, Y. (2011). Biocatalytic domino reaction: synthesis of 2H-1-benzopyran-2-one derivatives using alkaline protease from Bacillus licheniformis. Green Chemistry, Vol.13, No.8, (June 2011), pp. 2048-2054, ISSN 1463-9262

Wasserscheid, P. & Welton, T. (2002). Ionic Liquid in Synthesis, WILEY-VCH Verlag GmbH & Co. KGaA, ISBN 3-527-30515-7, Weinheim, Germany

Watson, T. J. N., Horgan, S. W., Shah, R. S., Farr, R. A., Schnettler, R. A., Nevill Jr, C. R., Weiberth, F. J., Huber, E. W., Baron, B. M., Webster, M. E., Mishra, R. K., Harrison, B. L., Nyce, P. L., Rand, C. L. & Goralski, C. T. (2000). Chemical Development of MDL 103371: An N-Methyl-D-Aspartate-Type Glycine Receptor Antagonist for the Treatment of Stroke. Organic Process Research & Development, Vol.4, No.6, (December 2000), pp. 477-487, ISSN 1083-6160

Watson, T. J. N., Horgan, S. W., Shah, R. S., Farr, R. A., Schnettler, R. A., Nevill Jr, C. R., Weiberth, F. J., Huber, E. W., Baron, B. M., Webster, M. E., Mishra, R. K., Harrison, B. L., Nyce, P. L., Rand, C. L. & Goralski, C. T. (2000). Chemical Development of MDL 103371: An N-Methyl-D-Aspartate-Type Glycine Receptor Antagonist for the

Treatment of Stroke. *Organic Process Research & Development,* Vol.4, No.6, (December 2000), pp. 477-487, ISSN 1083-6160

Xia, Y., Yang, Z., Brossi, A. & Lee, K. (1999). Asymmetric Solid-Phase Synthesis of (3′R,4′R)-Di-O-cis-acyl 3-Carboxyl Khellactones. *Organic Letters,* Vol.1, No.13, (December 1999), pp. 2113-2115, ISSN 1523-7060

Greenwashing and Cleaning

Develter Dirk[1] and Malaise Peter[2]
[1]Ecover Coordination Center NV,
[2]Meta Fellowship npo
Belgium

1. Introduction

The world population is rapidly growing: estimates tell us that from nearly seven billion actually, we're on our way to some eight to ten billion in 2050 (US Census Bureau, 2011). This ongoing growth generates a forever growing demand in products and services: raw materials, energy, transport and transformation capacity, waste disposal. Unfortunately, the backbone of all of these activities is mainly dependent of a single fossil source, crude oil and its derivatives. However, these have a double disadvantage:

- crude oil is only present in geographically limited areas of the planet
- the stock of reasonably accessible fossil matter is, after 180 years of industrial life, nearly depleted and cannot be replenished

Parallel to this huge growth and rapid depletion there is a growing consciousness on hygiene and personal deployment in all countries worldwide, but mainly in developing ones. The need for products and services is still growing exponentially.

Basically, this complex growth is already transcending the capacity of the planet, making the largest part of the average economic activities *unsustainable*. Some market segments already suffer from it, metals e.g.: prices for ores and metals have rocketed the last couple of years and the market for recycled materials is on an all time high. For some of them we are quite close to a shortage. It won't stop there: with decreasing mineral and fossil sourcing capabilities, producers will be forced to turn to non-fossil, non-mineral, renewable sources, that means in the first place plant sources and to a lesser extent, animal sources.

These raw material sources are actually and nearly exclusively the providers of food, but they will inevitably suffer a strong competition from non-food production demand. We already saw the consequences of unplanned and unregulated behavior on the matter when, in 2007, there was a sudden huge demand in renewable raw materials for the production of biofuel, the "Food vs. Fuel" crisis, also called the "Tortilla Flap". Tortilla prices doubled for the already poor Mexican population, causing riots. The real causes were not even a raw material shortage, but mainly speculations in the globalised markets (Kaufman, 2010; Nelson, 2008; Western Organisation of Resource Councils, 2007).

When one day, next to food, a large part of clothing, housing and utensils will forcedly have to be derived from plant and animal sources, there will not only be substantial shortages, high prices, fights and even wars to be expected in relation to those materials, but it will simply be impossible to generate such an amount of raw materials on this planet. At the actual consumption rate, available agricultural space is simply not big enough to provide all the necessary. The only reasonable outcome is a double action. We must at the same time substantially reduce our needs in raw materials and energy, and hugely increase research in new ways of raw material sourcing, higher efficiency and better transformation processes.

Such a type of development has already been proposed and documented by the Wuppertal Institute (http://www.wupperinst.org), Germany, under the names of "Factor Four" and "Factor Ten", targeting a fourfold, respectively tenfold reduction in the bulk of our needs. Their theme is still: *"We use resources as though we had four earths at our disposal"*, which describes the actual problem quite well. The Wuppertal Institute also developed instruments to quantify such developments, amongst them the *Material Input Per Service* tool (MIPS). This approach measures how much earthly substance is needed to generate all the necessary for one service of a given product.

As always with such developments we can discuss until Doomsday if this is the right thing to do here and now and if there are no better solutions (read: less compulsory, less drastic, financially cheaper etc.). Perhaps – but can we afford to wait? Do we really have the time to abide such ideal solutions? We, the authors, are convinced we don't and we prefer in this to marry the approach of the Swedish 'Natural Step' organization (http://www.naturalstep.org): *let's not be stuck on endless discussions about the twigs of problems, but let's agree about the trunk and the branches.*

2. Global pollution consequences

One of the main consequences of unsustainable economical activities is the continuous exposure of man to man-made chemicals, and to high levels of mixed, persistent pollution. Especially children, young people and the elderly are vulnerable. It's not that much the spectacular catastrophes, such as oil spills and sinking tankers, which are quite visible; but the silent, insidious spreading of chemicals that should not reside in nature at all. Heavy metals are notorious, but there are even more risky compounds, such as:

- pesticides and insecticides, for which Rachel Carson warned us as early as 1960 in her book *Silent Spring*
- PCB's and CFC's, as representatives of a huge family of chlorinated compounds
- preservation and disinfecting agents
- historical polluters set free by the meltdown of glaciers and polar caps

All of the consequences become only indirectly visible, through degrading biodiversity and fertility, making species disappear at an abnormal rate and speed, through high incidences of uncommon pathologies or even through the transmission of risky genetic properties. Unfortunately, economists and investors don't see these phenomena as relevant for economic life. It's true that when economic life is just looking at the next fiscal quarter, hardly anything is relevant. When our ancestors would have thought and acted like that, we wouldn't be here.

But there *is* a relationship with those phenomena - and a tight one too. One out of many immediate consequences of man-made pollution for example is the actual worldwide mortality amongst honey bees. Hardly noticed by most people but for their honey we consume, they are primarily responsible for pollination in nature. A declining pollination will, amongst others, have an immediate negative influence on agriculture and everything depending on it, but also on pollination in the wild. This will at the end turn the planet into an infertile, uneconomic desert.

3. Unsustainable processing

The production processes for most goods and services are mainly using unsustainable energy, involve unsustainable processing techniques, unsustainable transport methods and generate an unacceptable amount of problematic waste. Only very few countries have managed to switch a substantial part of their energy production to sustainable energy sources. Some of the actions that have been taken on the matter are questionable, because they only touch a small part of the problem. Saving bulbs are such an example: they leave about 81 % of the household consumption of electricity with fossil sources and make the consumer believe that he is solving a problem. Politically spoken the venue of saving bulbs was a quick and dirty decision, but in the meantime new health concerns have risen in relation to saving bulbs. Unfortunately, they are now becoming compulsory in many countries. We don't hear anything though on saving fridges, saving deep freezers, saving washing machines, saving dry tumblers, saving fryers, saving stoves; all of these devices are in their actual form the real culprits, counting for 81% of the energy consumption in households.

Saving cars exist to some extent: the hybrids are on sale since some years and some prototypes of electric cars are presented increasingly. However, their promotion is not taken serious enough, neither by their producers, nor by politicians. Most cars still have fuel consumption and CO_2 emission rates which are unacceptably high, although the improvement technology is available. Some car companies are known to actively lobby against stricter emission laws (Greenpeace, 2011). Public transport offers an interesting solution to a large part of the private traffic, but it is actually legging far behind as to comfort, frequency and efficiency. It will nevertheless be one of the main choices to realize a sustainable transport system for future society.

When we look at production processes the situation is even worse. Up to recently there was hardly any attention for sustainable industrial processing techniques. More or less as a rule such processes involve high temperature and high pressure, often accompanied by other energy demanding techniques such as vacuum generation, freezing or desiccation. The production and transformation of aluminum or the cracking of crude oil and the processing of many of their derivatives e.g., are such energy devouring activities. Some other sectors such as the production of chemicals for household and industrial uses can cause huge environmental and health problems. Hopefully we will not forget Bhopal (India), still an unsolved problem for the local victims, 20 years after the catastrophy occurred. Nor more recently the *Deepwater Horizon* (Mexico) and *Ganeth Alpha* (Aberdeen) ridge oil spills and the flood of poisonous aluminum sludge in the *Ajkai Timföldgyár* factory (Ajka, Hungary). These are just a few examples of the consequences of unsustainable production methods.

The production of commodities for the mass market is equally tainted. Detergents and their raw materials are for the essential part made from fossil sources, although they could easily be made from renewables – as a matter of fact they have been, up to about 1930. A large part of paper derived disposables still use freshly cut trees instead of recycled fibers. In many countries industries get an explicit advantage on their electricity bill because of their high consumption – whereas the opposite should be the case when we would apply the rule 'the pollutor pays'. There is hardly an incentive for companies to take serious action on the matter.

Production processes in the agricultural realm have, on top of high energy demands in some sectors, such as greenhouses, a huge impact on health and environment through the chemicals they introduce in the food web: synthetic fertilizer, insecticides and pesticides, as well as after-treatments for preservation and pest control. Potato culture knows an average of about 11 chemical treatments before harvesting. The production and use of banana pesticides and insecticides causes heavy health and environmental impacts (Chua, 2007). Soy production for animal feedstock, and palm oil production for food and non-food applications, are still devastating huge areas of virgin forest, destroying important natural CO_2 dumps.

The policy of subsidizing agricultural and other produce for export are still in place, even for such goods that can easily be produced in the destination countries. There is no sensible reason why Austrians should eat Belgian green peas instead of the ones from their own agriculture, and Belgians the Austrian ones, unless in case of a shortage on either side. Consuming as much as possible produce and products from where one lives could substantially reduce primary fuel consumption, traffic jam, pollution and health impacts. Farmers should be subsidized for maintaining the natural fertility of the soil and the preservation of biodiversity – not for overproduction within monocultures, as it is the case now. That would lead to a broad support and promotion of certified organic farming, rather than fighting it with prejudice. Organic farming has maintaining the natural fertility and preserving biodiversity in its basic principles – you can't have organic farming without them. The secondary effects of such measures would in the middle and long term be very important as well: no synthetic fertilizer, little to no chemical insecticides and pesticides, a healthy soil and a healthier water system.

The professional transport of food and non-food commodities knows comparable problems. But there is in the transport sector even less interest for sustainable transport solutions than with individuals: maximum load, high speed and low financial cost are the sole drivers. Some isolated projects, such as the one set up by the Belgian distributor Colruyt with an innovative lorry, goes further and tries to reduce fuel consumption and emissions by pro-actively financing, testing and adopting hybrid equipment that will respond to the newest EU requirements (Colruyt Group, 2010).

For any man made activity we should since long have adopted the *Precautionary Principle* (European Commission, 2000): when we don't know the consequences, or have difficulties in estimating the extent of health and environmental impacts – including the depletion of raw material sources – we just shouldn't do it.

4. Old stuff

These facts are not new. Rachel Carson wrote her book *Silent Spring* in 1960 (Carson,1960). She warned for a thoughtless, large scale use of man made, highly effective chemicals and documented the then already visible consequences for health and environment.

Starting in 1970, several high-level reports continued warning for the consequences of such an unsustainable development: *The Predicament of Mankind* (Christakis et al., 1970); *Limits to Growth* from the Club of Rome (Meadows et al., 1972); *Our Common Future* from the Brundtland Commission (World Commission on Environment and Development, 1987). But in spite of all these serious efforts and a series of follow-up initiatives such as the *Rio Conference, Agenda 21, Rio+10* and many more, very little systematic action has been taken.

The global principle to tackle what became a *global problem,* is *Sustainable Development.* The Brundtland report, in which the term 'sustainable development' was first used, describes this in a much cited quote as *"development that meets the needs of the present without compromising the ability of future generations to meet their own needs."* It has three focus points, which are inextricably intertwined and should not be separated at any time:

- a social focus
- an economical focus
- an ecological focus

It will be obvious that each of these three members has its own specific rules and laws, which might be influenced by the others, but not overruled or replaced by them. It is not possible that economic principles will become more important than social or environmental ones; but they cannot become less important either. An essential fact - often misunderstood even by fervent followers – is that Sustainable Development is a life style, not a status that one can reach some time. You can't be, or can't become 'sustainable', there will always be a further stage of development to attend.

However, to cut short any misunderstanding: when we will in the following write about 'sustainable raw materials' or 'sustainable energy' we are not pointing at a status those items are supposed to be in, but at the whole *process* that leads to their existence. The raw material is a crystallization point of a generative process which fits – or doesn't fit, or only partially fits – into principles of Sustainable Development.

It's also obvious that, on the short term, it will not be possible to realize all elements of Sustainable Development to an equal degree of fulfillment, immediately and at the same time. Many things will only be partially realized through compromises between societal partners, in a slow process of involvement and comprehending.

Sustainable Development encompasses sustainable design, sustainable raw materials, sustainable production processes, sustainable energy and services, green taxes, as well as sustainable consumption. In short, it's a *cradle-to-grave* approach at all times and a *cradle-to-cradle* approach whenever possible (more on this theme is to be found in Braungart and McDonough (2002). *Cradle-to-grave* means that all partners are part of the whole process, from the design of the product or service until the disposal of possible leftovers, and anything in between. Each step has to be optimized: it should involve the lowest amount of earthly substance and energy possible, have the highest efficiency and user friendliness possible and generate as little leftovers as feasible (that what we still call 'waste'). This should be featured without compromising elements such as the availability, the efficacy or the price of the product or service.

Cradle-to-cradle goes even a step further: whatever substance that is not fully destroyed at use (such as food), has to be made reusable for a similar, or even for a completely different application by similar or different producers. In doing so, 'waste' becomes non-existent, as it will be a raw material for a new process. This is the way nature acts, and nature is never short of raw materials, unless humans degrade its ways.

Can we secure such a development, such type of products and services, can we guarantee that this will work and that everything will be true and honest? No! One of the important elements to be redeveloped in parallel, is *trust*. Trust mustn't be blind, though; there are several mechanisms that can be put to work to coach Sustainable Development. *Green Taxes* are one of those, but they are sort of an end-of-pipe solution and they should preferably only be used as temporary, corrective measures. It makes no sense to implement such a huge beast as Sustainable Development by means of force. Another useful mechanism are *Green Labels*. We know a whole bunch of them all over the planet, they have since a couple of years grown like mushrooms, and not always for the good. Unfortunately these Green Labels have mostly been developed by an amalgamate of politics and industry, and we should not forget that this is the tandem that pushed us into *Unsustainability*. It's comparable to farmers and butchers deciding about the criteria for veggie burgers; that makes no sense, really.

When Green Labels have to play a proficient role - and we think they can and should - the consumers have to get far more grip on the process of green labeling, from designing over controlling to improving them. Politics can check their correct and equitable implementation, but should not decide on the content or the format. Industry has to listen to what the consumer wants, not enforce what they themselves want to produce and sell. When it goes like that, we end up with a mishap, such as the actual EU *Ecolabel* on detergents. It lacks all kinds of arms and legs: the raw material sourcing is evaluated on its ecological merits in a crippled way, there is no social element present in the model, product efficiency is stubbornly compared against conventional, thus unsustainable products (in other words: race car vs. bike). Very weird influencing from conventional industry circles crept in through doors and windows, and one of the consequences is that fragrances based on plants are in practice *prohibited* in ecolabeled products!

Public procurement is another extremely powerful mechanism to implement Sustainable Development. When all layers of public power should systematically include ecological and social criteria in their evaluations and tenders for products and services, and not just take the lowest price as a gauge, the usual product and service ranking might be turned upside down.

Companies and offices of all kinds can use the same strategy. New knowledge and understanding would be instilled slowly into society as a whole, because consumers would comprehend and follow the example.

Hovering back over what we wrote up to now, we can see that Sustainable Development is not just about some kind of environmental conservation, but clearly encompasses the economical, social and environmental issues we tried to describe.

5. Washing and cleaning

Why washing and cleaning? In the next two chapters we will mainly deal with ways to deploy, expand and improve the characteristics of Sustainable Development within commodity products and services, and the models used to do so. Both the authors have a longstanding experience in designing and modeling concepts, formulas and strategies for sustainable washing and cleaning products, as staff member and retired staff member with the worldwide market leader in the trade.

A traditional situation for commodities is that there is long trail of experience, starting in the past and ending today. But – as banks use to state lately – gains from the past are no guarantee for future gains. On the contrary; each and every conventional commodity is anchored in an unsustainable past and can by no means give reliable clues for a future that is headed by Sustainable Development. The trail for sustainable products and services has no past, it starts today and leads far into the future. Only, we don't know anything about that future. Keeping in mind what the Brundtland Report says: *"Sustainable Development is development that meets the needs of the present without compromising the ability of future generations to meet their own needs"* (World Commission on Environment and Development, 1987), we have the responsibility from now on to design products and services *in the wake of issues which lie in the future, in front of us!* It's a complete, revolutionary turning around of the way we used to think – and very challenging.

Products and services which have been conceived this way can be called *Future Capable*. Although not yet part of that future, they have the potential to fit in a future context. It will be clear that *unsustainable product design, unsustainable raw materials, unsustainable processing and energy, next to substantial waste generation, will altogether lead to a product that is not Future Capable.* That is a huge risk for the operations and investments of any company that would act in such a way. A producing company or developing lab will, on the contrary, try to the best of their knowledge and capabilities to become *Future Capable* as an organization.

But that is not easy and not immediately rewarding in terms of profit. Therefore, we will more often than not see some form of greenwashing popping up. There's all kinds of flavors, from smoothing some edges to sheer fraud, but always with the purpose to be perceived as "green". You have green petrol (it's toxic, carcinogenic and highly flammable), green apple perfume (green apples really don't have any perfume stuff) or natural soap (there is no soap tree on which that grows).

Unfortunately, many publications targeting a "green" consumer try to pick their grain as well. They start making product evaluations without having the technical knowledge to do so and without any real knowledge about environmental issues or Sustainable Development – except for the pure legal things, but those are always legging far behind reality. Thus is the consumer more or less left to himself in a no mans land.

This mustn't be, however. It is perfectly possible to select - or even develop - sustainable gauges. The starting point has to be to select or develop gauges for raw material sourcing, process technique selection and energy, product design, and finally the health and environmental impacts at use and after disposal. None of these can claim to be the ultimate complete tool, a "green meter", that gives you once and for all the mathematically exact ranking of whatever. Nature doesn't function like that, it's not a machine, and neither are we. But these tools can give us fairly reliable estimations.

Just two examples:

- The Eco-costs/Value Ratio (EVR) developed by the Technical University Delft, The Netherlands (http://www.ecocostsvalue.com).

- The Eco-Footprint, originally developed in Canada, in the meantime in use in different forms (http://www.myfootprint.org)

When held against such sustainable gauges, actual solutions can be evaluated and be put in a priority ranking for further improvement. Compromises will have to be made and a time frame accepted: not every critical element can be instantly replaced, for very different reasons, such as unavailability of materials or processing, technical incompatibilities or financial constraints – or all of them together. There are many examples from the recent past.

- Until recently, fridges used to function on chlorinated compounds (CFC's) which are ozone depleting, persistent, toxic and carcinogenic. They have been exchanged for one or two less risky compounds - which could have been done since long. Nevertheless, 'less risky' is not 'good' and other solutions have to be developed.
- Hybrid cars mainly use two different engines, a combustion one and an electric one. They have low consumption and low emissions. But hybrids are neither the solution for the mobility problem, nor are they the ultimate green cars; they just feature the Best Available Technology (BAT) of the moment.
- Ecosurfactants are a class of washing agents from renewable raw materials, made via fermentation, at low temperature, low pressure and zero waste. They outperform both petrochemical and plant based surfactants on efficiency. But not all needs of detergent concepts can yet be covered.
- In almost each country there are organizations which defend consumer interests. But they are more often than not axed on quite superficial, practical and price issues and hardly on sustainable ones. They mostly take a Calimero standpoint and don't really try to mediate between consumers and industry to develop a common ground.

Because of the complexity of the issues, each market segment will have to develop its own models and time frames. These models will need frequent revisions to fit in a forever changing context. All of us will have to learn to operate in a very different context. Where we are now in a closed circuit, with proprietary knowledge and confidentiality issues, we will have to adhere to Open Access, validation and control by external parties and sharing of know-how. It seems that the idea of competition in the old sense is getting quite rusty in this changing world and asks rather for models based on communication and collaboration.

In the next chapter, we describe the elements and the backbone of such a model for washing and cleaning commodities as developed and used by Ecover, based on the above ideas.

6. Ecover's "diamond" model

Ecover is a medium sized, Belgium based company and is one of the foremost pioneers in developing and manufacturing washing and cleaning products with respect for the environment. It started off three decades ago by deleting environmentally troublesome ingredients (such as phosphates, alkyl phenol ethoxylates and the like) from standard frame formulas. This resulted in the so-called "No-code" (product doesn't contain such and doesn't contain so), which was communicated on the packaging.

Oleochemical based alternatives to petrochemicals were used wherever possible – there were not very many, 30 years ago. This black-or-white approach, though easy to understand by the consumer, did by no means automatically guarantee a satisfactory product performance. But such was the understanding of the post-hippie generation: a product was considered "environmental", or even worse, "natural" when it did not contain certain ingredients which were on a relatively vague blacklist. Over the years and in the wake of the appearance of a forever growing number of renewable raw materials, it was replaced by a more pragmatic approach, the in the meantime quite well known and respected "Ecover Concept".

Today, Ecover's environmental product profile is maximized to achieve market standard performance. If the balance of a basic set of criteria (price, performance, convenience, human safety, environmental profile) of a functional ingredient, is considered prone to improvement, an ingredient development project is set up, usually in cooperation with academic and/or industrial partners. This approach necessitates a quantitative tool to measure ingredient and product strengths and weaknesses and to allow a company wide evaluation of innovation progress. Ironically enough, a large part of the market where Ecover acquired sales strength over the last decade is reluctant to even try to understand the full story; yet they desire to make the best environmental purchase. They are looking for a simple approach or even some authority who can tell them what is right and what is wrong; as we learned however, there is no such situation in the real world. But some ecolabel schemes deliver exactly this pass/fail endorsement, without necessarily featuring a coherent product picture.

The challenge for Ecover therefore was to develop a model based on externally verifiable data, encompassing the largest part of the Ecover concept, yet easy to understand for the non-chemist and allowing almost instant appreciation of a product's profile, both within the Ecover company and among its consumers. By furthermore incorporating European ecolabel criteria into the model and having this model validated and controlled on a yearly basis by an independent third party (in this case the Belgian NPO *Vinçotte Environnement*) the Ecover "diamond" model (fig.1) has become a strong and difficult to dispute communication tool. It takes Ecolabel criteria to a higher level by incorporating criteria on ingredient sourcing and adhering to stricter standards with regard to environmental impact and leftover fate. The "Ecover diamond" model (thus named because of the diamond-like structure of its visualisation) can be considered as a self-declared environmental claim according to ISO 14021. Self-declarations are more often than not unreliable and untrustworthy, but here we have one that responds to strict external regulations and controls. An important requirement for environmental claims and their evaluation is their scientific basis, with only clearly referenced methods, calculations and standards.

The diamond model has proven to be a useful tool in product development, product benchmarking (comparing Ecover products with market references) and in communication. The Ecover diamond is the methodological translation of most of the Ecover environmental concept as it has been around for more than a decade. The procedure describing the diamond compilation also includes "focal drivers" mainly pertaining to qualitative criteria and to criteria which are hard if not impossible to assess for competing products. These focal drivers thus embody additional Ecover commitments not visualised in the diamond and often not communicated in any way. In this respect the diamond procedure has in fact become the written compilation of the Ecover concept.

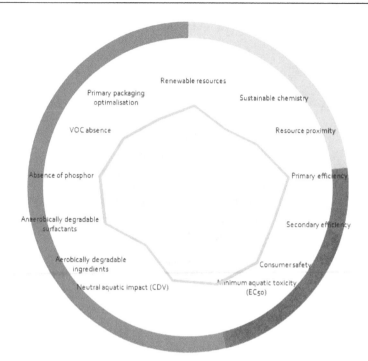

Fig. 1. Ecover diamond model with 13 axes distributed over 3 life cycle phases.

The model involves the total life cycle of a product, the **Extraction Phase**, the **Usage Phase** and the **Absorption Phase**. The latter phase is termed "Absorption" rather than the standard LCA "Disposal phase" terminology, to refer to a cradle-to-cradle, closed carbon loop, without persistent chemicals and without lasting ecosystem perturbation. It is visualized as a spidergram with 13 quantitative axes distributed over the three said phases.

The **Extraction Phase** involves *Renewable Resources*, *Green Chemistry* and *Material Proximity*. *Renewable Resources* are defined as animal, vegetable or microbial derived feedstocks, as opposed to water, mineral and petrochemical resources. This axis represents the percentage of renewable matter over the total organic dry matter of the end product and correlates very good with experimental C_{14} carbon dating results.

The *Green Chemistry* axis reflects Ecover's striving for efficient resource transformation at low temperature and pressure, without potential run away reactions or risk of explosion, while making use of chemicals currently considered as safe, with limited risk of undesirable by-product formation. The axis is calculated by a weighted sum of "green chemistry scores" across all ingredients over the total organic dry matter.

Resource Proximity covers the CO_2 contribution of the complete product formula, from the source of the ingredient constituents, to the ingredient manufacturer, to the Ecover factory in Malle, taking into account the distance traveled by all individual ingredients and their transport mode.

The **Usage Phase** involves Primary Efficiency, Secondary Efficiency and Consumer Safety.

The *Primary Efficiency* is the immediately perceivable performance of a product. This axis represents the percentage of "performance score", relative to a reference formula and determined according to EU Ecolabel standards.

Secondary Efficiency is the performance of the product at lower temperature (in automated appliance products) or a second performance attribute (such as speed of drying, gloss retention, …), again relative to a reference formula.

Consumer Safety covers the use of surfactants that are safe for the user. Several attributing points towards consumer safety are defined. The absence of certain danger classes (e.g. corrosive, toxic, …) attributes a percentage to the total score.

The **Absorption Phase** involves Aquatic Safety, Limited Aquatic Impact, Aerobically Degradable Ingredients, Anaerobically Degradable Surfactants, Phosphorus Absence, VOC Absence and Primary Packaging Optimisation.

Aquatic Safety covers the use of ingredients that are safe for the aquatic environment and is determined experimentally at Ecover as aquatic toxicity tests and expressed as a dose related LC_{50} quotient.

Limited Aquatic Impact is calculated as the Critical Dilution Volume (CDV), a concept developed within the EU ecolabel and expressing the theoretical amount of liters required to dilute a single product dose down to environmentally harmless concentrations, provided sewage treatment systems are in place.

Aerobic Biodegradability is an important and desirable property of any ingredient in washing and cleaning products. This diamond axis visualizes the amount of persistent chemicals in the product, i.e. the chemicals that are not inherently degradable by microorganisms when oxygen is present. An ingredient can be readily biodegradable, inherently biodegradable or persistent in aerobic conditions. This clearly differentiates the diamond model from ecolabel criteria.

Anaerobically Degradable Surfactants excludes surfactants which are not biodegradable in anaerobic conditions, i.e. in oxygen deprived environments should be avoided to the extent possible since aerobic conditions are not always the case, such as in many rivers, marine sediments or sewage sludge.

Phosphorus Absence documents possible amounts of phosphorus, which in the aquatic environment causes eutrophication. Hence, the use of phosphorus-based ingredients should be minimized.

The environmental relevance of *Volatile Organic Carbons* (VOC) is the contribution to indoor air pollution and smog formation.

The *Primary Packaging* axis aims at reducing this waste according to several references and assumptions.

For more detailed information on the Diamond Model, see at www.ecover.com (or specific URL).

7. References

Braungart M.& McDonough, B. (2002). Cradle to Cradle: Remaking the Way We Make Things, North Point Press, New York.

Carson, R. (1960). Silent Spring, First Mariner Books, ISBN 0-618-24906-0, edition 2002

Christakis, H.; Jantsch, E.& Özbekhan, H. (1970). The Predicament of Mankind, Date of access 27/10/11, Available from:
http://sunsite.utk.edu/FINS/loversofdemocracy/Predicament.PTI.pdf

Chua, J. (2007). Latin American Banana farmers sue over pesticides. In: TreeHugger, 27/10/11, Available from:
http://www.treehugger.com/files/2007/08/pesticide_lawsuit.php

Colruyt Group, 2010. Groep Colruyt ontwikkelt hybride trekker, press release 21/06/2010, Available from: http://www.colruytgroup.be/colruytgroup/static/energiebeleid-hybride_be-nl.shtml

European Commission, 2000. Communication from the Commission of 2 February 2000 on the precautionary principle, Available from:
http://europa.eu/legislation_summaries/consumers/consumer_safety/l32042_en.htm

Greenpeace, 2011. Turn VW away from the Dark Side. Press Campaign, Available from:
http://www.vwdarkside.com/en

Kaufman F. (2010). The Food Bubble: How Wall Street starved millions and got away with it, Harper's Magazine, July 23, 2010.

Meadows, D.; Meadows, D.; Randers, J.& Behrens III, W. (1972). The Limits to Growth. Universe Books , ISBN 0-87663-165-0, New York:

Nelson, S. (2008). Ethanol no longer seen as big driver of food price, Reuters Press Release 23/10/08, Available from: http://uk.reuters.com/article/2008/10/23/food-corn-ethanol-idUKN2338007820081023

US Census Bureau, 2011. World POPClock Projection, Available from :
http://www.census.gov/population/popclockworld.html

Western Organisation of Resource Councils (WORC), 2007. Fact sheet October 2007

World Commission on Environment and Development, 1987. Our Common Future, Report of the World Commission on Environment and Development, Published as Annex to General Assembly document A/42/427, Development and International Co-operation: Environment, Available from:
http://worldinbalance.net/intagreements/1987-brundtland.php

3

Electrochemically-Driven and Green Conversion of SO$_2$ to NaHSO$_4$ in Aqueous Solution

Hong Liu[1,2,*], Chuan Wang[1,2] and Yuan Liu[1]
[1]*Chongqing Institute of Green and Intelligent Technology,*
Chinese Academy of Sciences, Chongqing,
[2]*School of Chemistry and Chemical engineering,*
Sun Yat-sen University, Guangzhou,
China

1. Introduction

The world has widely resorted to fossil fuels to power the industry and everyday life. In China, above 70% of the energy is extracted from coal. Emission of SO$_2$ due to burning of fossil fuels, in particularly of coal, causes harmful impacts on the environment, human health, livestock, and plants *(1,2)*. Many measures have been taken to cut off the emission of SO$_2$ during the last generations. It can be seen that the reduction of SO$_2$ emission in developed countries such as the United States has been witnessed *(3)*. However, due to the aspiration of energy to drive the economic increase and sustain the expanded population, the SO$_2$ emission is estimated to augment sharply in the rapidly developing Asian areas, and will still pose as a worldwide environmental problem in the next 30 years *(4)*.

Basically, the SO$_2$ emission is reduced after the burning processes through various flue gas desulfurization (FGD) processes *(5-8)*, which serves to transform the S(IV) to S(VI) and frequently to immobilize the SO$_2$ waste in the form of a solid. Of them, a wet limestone FGD process *(6,9,10)* using CaCO$_3$ mineral, represents over 90% of the installed desulfurization capacities in the world *(9)*, and can be chemically expressed below :

$$SO_2 + CaCO_3 + 2H_2O + \frac{1}{2}O_2 \rightarrow CaSO_4 \cdot 2H_2O + CO_2 \qquad (1)$$

Eq 1 illustrates that the SO$_2$ is transformed and immobilized in the form of CaSO$_4$·2H$_2$O, which may be commercialized as gypsum, but the incentive is little in areas including the United Sates *(11)* and China because of its rich natural sources. To our knowledge, only 3% of the FGD byproduct gypsum can be reused in China. In fact, once treated improperly, the solid waste becomes a secondary pollutant, and thereby is of a great environmental concern. Meanwhile, eq 1 shows that 1 mole of SO$_2$ leads to 1 mole of CO$_2$, whose discharge and

* Corresponding Author

accumulation in the atmosphere is recognized to aggravate a greenhouse effect *(12,13)*. Actually, it is considered that most wet FGD processes have an inherent shortcoming of secondary pollution, or of high running cost if the secondary pollutant is avoided. A challenge to overcome such shortcomings still remains *(14)*.

To meet the challenge, novel green technologies with no/less secondary pollution and with a value-added product become essential. Fan et al. have developed a process of converting the SO_2 to polymeric ferric sulfate, which can be employed as a common coagulant for water and wastewater treatment *(12)*.

Electrochemical techniques, utilizing electrons as a clean reagent, exactly enjoy the sustainability. Since most wet FGD processes embrace a sub-process of electron transfer for the oxidation of S(IV) to S(VI), the electrochemical techniques appear to fitfully work there. The electrochemical cleanup of flue gas has already been tested. For example, SO_2 can be anodically oxidized to H_2SO_4 in aqueous solution *(14-16)*, and regeneration of FGD agents is developed by using electrodialysis with a bipolar membrane *(17,18)*. It should be noted that as air coexists with the SO_2 in flue gas, electrochemical utilization of the molecular oxygen from air to further oxide the SO_2 is indispensable and should be encouraged. Such a new concept, however, has not been implemented so far.

To convert the SO_2 to be a value-added product without secondary pollution, this study aims at developing such a novel and green process by designing a series of electrochemical reactions through a SO_2 absorption-and-conversion process. In the process, a few considerations in the process can benefit the attempt. (i) The cathodic reaction utilizes O_2 from air to scavenge the process-released H^+ ions, while the anodic reaction uses H_2O to supply H^+ ions. (ii) The H^+ scavenging benefits the SO_2 absorption and its further oxidation. (iii) The H^+ supply benefits the formation of a bisulfate. Consequently, the SO_2 conversion is driven electrochemically to form $NaHSO_4$ as a sulfur-containing product.

$NaHSO_4$ is a valuable chemical and widely used as an additive in manufacture of dye stuff, a soil amender in agriculture, and replacement of H_2SO_4 in industry for pH adjustment and catalytic reactions. This study focused on the chemical and sustainable fundamentals as well as the pH optimization for the SO_2 oxidation. The findings are expected to lay a basis of understanding this new design with potential to convert the SO_2 from flue gas to $NaHSO_4$ as a value-added product in a green way.

2. Chemical fundamentals of the process

In this process, the SO_2 is designed to be absorbed into aqueous solution with alkaline, then oxidized to sulfate, and then transformed into bisulfate. The chemical fundamentals should be clarified to understand how the process works.

Absorption of SO_2. In the wet FGD processes, H^+ ions are released upon the absorption of SO_2 into the aqueous solution *(5)*:

$$SO_2 + H_2O \Leftrightarrow H^+ + HSO_3^- \tag{2}$$

$$HSO_3^- \Leftrightarrow H^+ + SO_3^{2-} \tag{3}$$

Oxidation of SO_2 to SO_4^{2-}. After the absorption, the absorbed SO_2 is oxidized by air in aqueous solution at moderate pH from HSO_3^- and SO_3^{2-} to SO_4^{2-} ions. This oxidation process is expressed as follows, of which eq 4 releases H^+ ions, but eq 5 does not *(19)*:

$$HSO_3^- + \frac{1}{2}O_2 \rightarrow SO_4^{2-} + H^+ \tag{4}$$

$$SO_3^{2-} + \frac{1}{2}O_2 \rightarrow SO_4^{2-} \tag{5}$$

ESH. The H^+ ions released through eqs 2~4 need to be scavenged due to their hindrance of the SO_2 absorption. Otherwise, the continuous absorption of SO_2 will be terminated. It can be noted that while O_2 in air is utilized for the SO_2 oxidation through eqs 4 and 5, the cathodic reduction of O_2 can be employed to scavenge the process-released H^+ ions. The reactions of O_2 reduction through a 2-electron process at acidic and neutral/alkaline conditions are expressed in eqs 6 and 6', respectively *(20-22)*:

$$O_2 + 2H^+ + 2e \rightarrow H_2O_2 \tag{6}$$

$$O_2 + 2H_2O + 2e \rightarrow H_2O_2 + 2OH^- \tag{6'}$$

At the same time, a side reaction co-exists with eq 6 below:

$$2H^+ + 2e \rightarrow H_2 \uparrow \tag{7}$$

It can be seen that eqs 6 and 7 consume H^+ ions and eq 6' supplies OH^- ions. All these reactions can be utilized to scavenge the H^+ ions released through eqs 2~4.

Transformation of SO_4^{2-} to Bisulfate. As coupled to the cathodic reactions, an anodic reaction is shown below:

$$H_2O \rightarrow \frac{1}{2}O_2 \uparrow + 2H^+ + 2e \tag{8}$$

Under an extremely acidic condition, the H^+ ions in eq 8 are combined with the SO_4^{2-} ions formed in eqs 4 and 5 to form a bisulfate:

$$SO_4^{2-} + H^+ \Leftrightarrow bisulfate\ ion \tag{9}$$

As a result, a model experiment becomes necessary to chemically substantiate this design by disclosing the ESH effect on the SO_2 absorption and oxidation, and by confirming the formation of $NaHSO_4$ as a product of desulfurization.

3. Experimental section

Chemicals and Reagents. SO_2 gas (99.9%) was obtained from KEDI, Foshan, China. Other chemicals as analytical reagents were used as obtained. Double distilled water was used in all experiments.

Experimental Procedure. The model experiment to substantiate this design was performed in an experimental setup as schemed in Figure 1, which consists of two 200 mL chambers: a DS chamber with a graphite rod (ø = 6.4 mm, and L = 200 mm, Chenhua, Shanghai, China) as the cathode and a saturated calomel electrode as the reference electrode, and an SR chamber with a Pt flake (2 × 1.5 cm², Chenhua, Shanghai, China) as the anode. Both chambers were connected by a salt bridge containing saturated Na_2SO_4 solution with agar. A PS-1 potentiostat/galvanostat (Zhongfu, Beijing, China) was employed to apply a cathodic current. The solution temperature was kept by a water bath at 25.0 ± 0.5 ºC and monitored by a thermometer. Air was purged onto the cathode surface by an air pump through a glass frit diffuser, and a needle valve was used to control its flow rate.

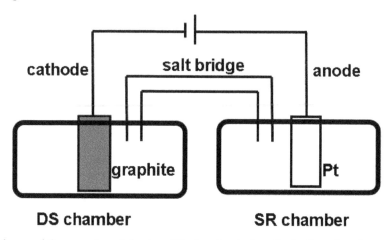

Fig. 1. Scheme of the experimental setup: DS chamber is desulfurization chamber, and SR chamber is sulfur-recovery chamber.

To quantify the OH^- ions electrochemically generated in the DS chamber and the H^+ ions in the SR chamber, electrochemical reactions of eqs 6, 6', 7 and 8 were performed in 0.01 M Na_2SO_4 solution without SO_2 at pH_0 6.0, in which an air flow of 100 mL min⁻¹ and different current densities of 0.10, 0.15, 0.20, 0.25, and 0.30 mA cm⁻² were applied. During the reactions, 0.01 M HCl and 0.01 M NaOH solutions were fed by a pump (Longer BTOQ-50M, Baoding, China) into the DS chamber and SR chamber, respectively, to maintain the pH at 6.0 ± 0.2. At the same time, 1.0 mL of solution sample was taken from the DS chamber for the quantification of H_2O_2 generation.

After that, two steps of the model experiment were performed in a batch mode. Step I was performed in the DS chamber, and Step II, in the SR chamber. Each experiment was performed three times and the values of experimental data in average are presented.

Prior to Step I, a start-up procedure was carried out to pre-store OH^- ions in the DS chamber through eqs 6', and thus a solution $pH_0 \geq 9.0$ in this chamber was obtained. The start-up procedure was described below, which was performed in the setup as schemed in Figure 1. Upon application of a cathodic current density as large as 0.60 mA cm⁻¹ to preclude the generation of cathodic byproduct H_2O_2 (20), air with 100 mL min⁻¹ flow rate was bubbled into water in the DS chamber to allow eq 6' to occur. The coupled reaction of eq 8 occurred simultaneously in the SR chamber. This start-up procedure continued till the pH increase to ≥ 9.0 in the DS chamber. Then, the solution in the SR chamber was discarded, and the solution in the DS chamber containing the pre-stored OH^- ions was utilized to absorb the SO_2 gas in Step I of the model experiment.

Thereafter, the SO_2 absorption of Step I was performed without current application. Initially, nitrogen gas was bubbled into the solution to remove any oxygen, then gaseous SO_2 was introduced into the alkaline solution to form $A-SO_2$ till pH decreased below 7.0.

In the $A-SO_2$ oxidation of Step I, an air flow at 100 mL min^{-1} was purged onto the cathode placed in the $A-SO_2$ solution. And a cathodic current was applied to maintain the electrochemical reactions. The $A-SO_2$ oxidation proceeded till the solution pH recovered to neutral pH (7.0).

To optimize the pH for the $A-SO_2$ oxidation, a set of experiments was performed at 1.0 mM $A-SO_2$ concentration, and different pHs in the range of 4.0~8.0 were maintained by chemical dosing of 0.01 M NaOH solution except at pH_0 8.0.

In the transformation of SO_4^{2-} to bisulfate of Step II, the reacted solution of Step I was relocated from the DS chamber to the SR chamber. Step II proceeded under a cathodic current with solution pH decrease in the SR chamber, while it stopped upon that the pH in the DS chamber reached the pH_0 value at the start-up step.

In Step I, the $A-SO_2$ concentrations were monitored by taking 1.0 mL of solution samples at pre-set time intervals, and 5 µL of methanol was injected into the samples taken during the $A-SO_2$ oxidation to quench any possible radical reaction (23). After Step I, SO_4^{2-} concentrations were measured. After Step II, H^+, Na^+, and SO_4^{2-} concentrations were measured.

Notably, in actual wet FGD processes, the SO_2 absorption and oxidation in Step I occur concurrently. However, to understand the ESH effect on the SO_2 absorption and its oxidation independently, the two experiments were conducted separately. At the same time, the air content in actual FGD processes should be minimized, and thus a small rate of 100 mL min^{-1} was fixed without further optimization. It was believed that an alkaline condition at pH > 7.0 was beneficial to the SO_2 absorption, and an acidic condition at pH < 3.0 was beneficial to the $NaHSO_4$ formation, while the $A-SO_2$ oxidation relied on pH, so the pH optimization for the $A-SO_2$ oxidation was performed.

Chemical Analysis. Measurement of the concentrations of electrochemically-generated OH^- and H^+ ions is accomplished by counting the dose of added H^+ solution with a known concentration to determine the amount of OH^-, and by counting that of OH^- to determine that of H^+. The $A-SO_2$ concentration was determined in terms of S(IV) concentration using a UV-VIS spectrosphotometer (TU1810, Universal Analysis, Beijing, China) according to a reported procedure (24). The H_2O_2 concentration was determined by spectrophotometry according to a potassium titanium (IV) oxalate method (20). The SO_4^{2-} and Na^+ concentrations were measured by ion chromatography (Dionex DX-600, U.S.). The cyclic voltammetry of $A-SO_2$ solution was performed in a N_2-saturated solution with 0.05 mM Na_2SO_4 as electrolyte on a CHI work station (Chenhua, Shanghai, China) with 50 mV s^{-1} scanning rate. The pH was monitored by a pH meter (PB-10, Sartorius, Shanghai, China).

4. Results and discussion

Electrochemical Generation of OH- and H+ Ions. This designed process underlined (i) that the electrochemical scavenging of process-released H+ ions due to eqs 2~4 would benefit the SO_2 absorption and oxidation, and (ii) that the electrochemical supply of H+ ions through eq 8 would realize the $NaHSO_4$ formation. Three cathodic reactions of eqs 6, 6′, and 7 served to increase the solution pH and electrochemical generation of OH- ions should be considered. Figure 2A illustrated that the OH- concentration in the DS chamber increased proportionally to the reaction time and the accumulative rate of OH- ions depended on the applied current density. In the meantime, H_2O_2 was generated through eq 6 and 6′. Figure 1B revealed that the H_2O_2 concentration increased against the reaction time. As paired to the cathodic reactions, anodic reaction of eq 8 occurred to supply H+ ions in the SR chamber. Measurements of the H+ ions revealed that the H+ concentration increased at the same rate as that of OH- ions in the DS chamber (not shown here).

The electrochemically-generated OH- ions in the DS chamber and H+ ions in the SR chamber carried electrons, which must be balanced electrically to keep the electrical neutralization of solution in each chamber. An analysis of electron balance will be described later.

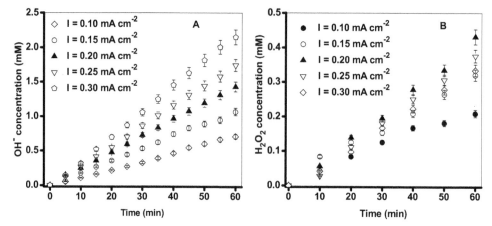

Fig. 2. Buildup of electrochemically-generated OH- ions (A), and H_2O_2 (B) in the DS chamber.

ESH Effect on SO_2 Absorption. The above electrochemically-generated OH- ions could be utilized to scavenge the absorption-released H+ ions through eqs 2 and 3. ESH has two functions. One was the pre-storage of OH- ions through eq 6′, which later served to scavenge the SO_2-absorption-released H+ ions. The other was the *in situ* scavenging of the process-released H+ ions through eqs 6 and 7 or *in situ* supplying OH- ions through eq 6′, which also served to scavenge the A-SO_2-oxidation-released H+ ions.

To disclose the ESH effect on the SO_2 absorption, one SO_2 absorption experiment in Step I of the model experiment was performed at pH_0 9.0 with pre-stored OH- ions, and then two additional SO_2 absorption experiments without pre-stored OH- ions were performed at pH_0 5.0 and 7.0, respectively. Figure 3 revealed that under the three pH conditions, the A-SO_2

increased first quickly, then slowly to a plateau. Comparatively, at pH_0 9.0, the A-SO₂ concentration increased most rapidly and ended at the highest level. The acceleration of SO₂ absorption at pH_0 9.0 was caused by the lifted pH, which was realized by pre-storing the OH⁻ ions electrochemically. Obviously, the pre-stored OH⁻ ions served to scavenge the SO₂-absorption-released H⁺ ions as a means of ESH. Thus, the SO₂ absorption was accelerated by the ESH.

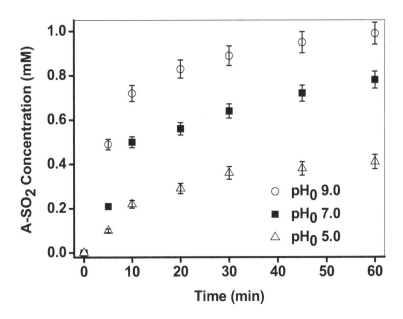

Fig. 3. Buildup of A-SO₂ concentration in the SO₂ absorption at solution pH_0 5.0, 7.0, and 9.0.

ESH Effect on A-SO₂ Oxidation. The oxidation-released H⁺ ions were in situ scavenged through eqs 6, 6′ or 7. To disclose the ESH effect on the A-SO₂ oxidation, two sets of SO₂ oxidation experiments in Step I were carried out. In the first set with ESH, a current density at 0.20 mA cm⁻² was applied to maintain the electrochemical reactions, while in the second set without SH, no current density was applied.

Figure 4 showed that with ESH, 100% and 95.8% of A-SO₂ disappearances were achieved at 30 min for the 1.0 mM A-SO₂ and 1.5 mM A-SO₂ solutions, respectively. Following the oxidation reaction, SO_4^{2-} concentrations were detected, and the results as listed in Table 1 indicated that 95.0% and 88.0% of the A-SO₂ were converted to SO_4^{2-} ions. By contrast without SH, only 70.7% of the 1.0 mM A-SO₂ and 60.9% of the 1.5 mM A-SO₂ disappeared after 30 min, and the SO_4^{2-} concentrations in the reacted solution were significantly lower than those with ESH (Table 1). From these it could be understood that the A-SO₂ oxidation with ESH proceeded more rapidly than that without SH, since ESH was beneficial to the conversion of A-SO₂ oxidation to SO_4^{2-}.

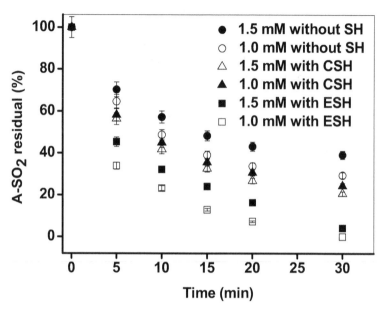

Fig. 4. Temporal disappearance of A-SO$_2$ concentration in the A-SO$_2$ oxidation at pH$_0$ 6.0, with ESH at 0.20 mA cm^{-2} current density for 1.0 mM A-SO$_2$ and 0.25 mA cm^{-2} current density for 1.5 mM A-SO$_2$, with CSH, and without SH.

A-SO$_2$ (mM)	SO$_4^{2-}$ with ESH (mM)	SO$_4^{2-}$ with CSH (mM)	SO$_4^{2-}$ without SH (mM)
1.0	0.98	0.78	0.66
1.5	1.32	1.15	0.97

Table 1. SO$_4^{2-}$ concentrations after the A-SO$_2$ oxidation at pH$_0$ 6.0, with ESH at 0.20 mA cm^{-2} current density for 1.0 mM A-SO$_2$ and 0.25 mA cm^{-2} current density for 1.5 mM A-SO$_2$, with CSH, and without SH.

To further disclose the ESH effect, a set of experiments in 1.0 mM and 1.5 mM A-SO$_2$ solutions was carried out with CSH, and the results are added in Figure 4. Clearly, the A-SO$_2$ oxidations with CSH proceeded more rapidly than those without SH, but more slowly than those with ESH. Therefore, it was further confirmed that the SH benefited the A-SO$_2$ oxidation, while the ESH was more effective than the CSH.

Moreover, the results in Table 1 showed that more SO$_4^{2-}$ ions were obtained in the reacted solution with ESH than those with CSH. It was believed that the H$_2$O$_2$ produced through eq 6 or 6' could enhance the SO$_2$ oxidation as an oxidizing reagent (27,28). After the reaction, no H$_2$O$_2$ residue left as impurity in the final sulfur-containing product. Therefore, the ESH benefited the A-SO$_2$ oxidation with two advantages of (i) scavenging the absorption- and oxidation-released H$^+$ ions, and (ii) simultaneously generating H$_2$O$_2$ to facilitate the conversion of A-SO$_2$ to SO$_4^{2-}$.

Transformation of Na_2SO_4 to $NaHSO_4$. Beyond the utilization of cathodic reaction to scavenge the H^+ ions released in eqs 2~4, this new process utilized an anodic reaction (eq 8) to supply H^+ ions which combined the SO_4^{2-} to form bisulfate. In both Steps I and II, the H^+ ions were produced in the SR chamber. Step II was performed by relocating the reacted solution of Step I from the DS chamber to the SR chamber where eq 8 occurred. Thus after this step, three ions of Na^+, H^+, and SO_4^{2-} presented in the SR chamber, and the analysis of mass balance of electrons as would be shown later suggested that a mixture of the three ions might result in the formation of $NaHSO_4$ through eq 9.

To confirm the $NaHSO_4$ formation, after the two-step model experiments with 1.0 mM and 1.5 mM A-SO_2 concentrations, the concentrations of the three ions in the SR chamber were measured. The results as shown in Figure 5 demonstrated that a mass balance of Na^+, H^+ and SO_4^{2-} ions was approximately 1:1:1, which ensured the $NaHSO_4$ formation.

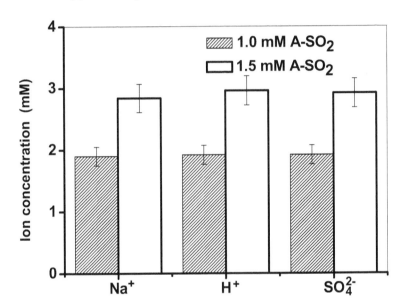

Fig. 5. Ion concentrations measured in the SR chamber after the model experiment.

Optimization of pH for A-SO_2 Oxidation with ESH. Figure 6A demonstrated that pH 5.0~6.0 was optimal for the A-SO_2 oxidation, which was in good agreement with others' results of optimal pH 6.0 kept by addition of OH^- ions in the SO_2 oxidation by air during a seawater FGD process *(24)*. Figure 6B showed the records of cyclic voltammetry in A-SO_2 solution, which confirmed that the optimal pH for the SO_2 oxidation was 5.0~6.0.

Figure 6B illustrates the cyclic voltammetry recorded in the 1.0 mM A-SO_2 solution, and the peaks at 0.12~0.15 V and 0.81~0.83 V were associated with the HSO_3^- oxidation and the SO_3^{2-} oxidation, respectively. On the other hand, the A-SO_2 solution consists of two major species of HSO_3^- and SO_3^{2-} in the pH range of 4.0~9.5. It has been documented that the HSO_3^- species occupies 100%, 92%, 38% and 10% of A-SO_2 at pH_0 5.0, 6.0, 7.0 and 8.0, respectively *(27)*.

Clearly, the peaks associated with the HSO$_3^-$ oxidation were not pronounced at pH 7.0 and 8.0, while the peaks were mature at pH 5.0 and 6.0. On the contrary, the peaks associated with the SO$_3^{2-}$ oxidation were mature at pH 7.0 and 8.0, while no peak was observed at 5.0 and 6.0 since the SO$_3^{2-}$ species were only 8% and 0, respectively. These results implied that at optimal pH 5.0~6.0, the HSO$_3^-$ predominated in the A-SO$_2$. Moreover, the HSO$_3^-$ oxidation proceeded more rapidly at this pH range than at pHs beyond this range. Thus, the optimal pH for the A-SO$_2$ oxidation was pH 5.0~6.0.

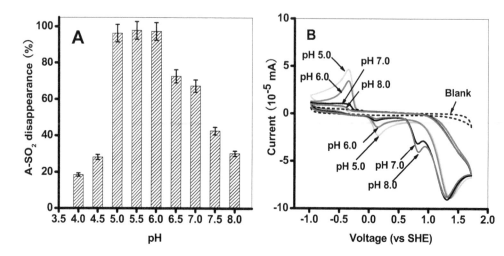

Fig. 6. Dependence of 1.0 mM A-SO$_2$ disappearance after 30 min oxidation on the solution pH (A), and cyclic voltammetry recorded in 1.0 mM A-SO$_2$ solution (B).

In real application, the current density should be adjusted to maintain the optimal range of pH 5.0~6.0 for the A-SO$_2$ oxidation. Consequently, a set of 1.0 mM A-SO$_2$ oxidation experiments was performed at pH$_0$ 5.0, 6.0, 7.0, and 8.0. It was found that when the current densities were adjusted to 0.20, 0.25, and 0.30 mA cm^{-2}, the optimal pH 5.0~6.0 could be maintained for pH$_0$ 7.0, 6.0, and 5.0, respectively. Evidently, the current density to maintain the optimal pH increased with the pH$_0$ decrease of A-SO$_2$ solution. The lower pH$_0$ meant more H$^+$ ions in the solution, so the higher current density was required to scavenge the H$^+$ ions. At pH$_0$ 8.0, the SO$_3^{2-}$ species predominated in the A-SO$_2$ solution (27), and no pH decrease was observed in the A-SO$_2$ oxidation by air as indicated by eq 5. Thus no application of cathodic current to generate OH$^-$ ions was required.

Analysis of Electron Balance in DS Chamber and SR Chamber. In this study, OH$^-$ ions were generated through eq 6' in the DS chamber, and H$^+$ ions were generated through eq 8 in the SR chamber. The electrons carried by the OH$^-$, H$^+$, and other emerged ions must be balanced electrically to keep the electrical neutralization of solution in either chamber. Table 2 lists the ion species after each step in the DS chamber and SR chamber. From the ion species, the analysis of electron balance could be made.

	DS chamber	SR chamber
Step I	HSO_3^- Na^+ (pre-stored) SO_4^{2-} from eq 4 Na^+ from salt bridge OH^- from eq 6'	H^+ from eq 8 SO_4^{2-} from salt bridge
Step II	OH^- from eq 6' Na^+ from salt bridge ----	H^+ and SO_4^{2-} coming from Step I Na^+ and SO_4^{2-} relocated from DS chamber of Step I, H^+ from eq 8 SO_4^{2-} from salt bridge

Table 2. Ion species after each step in the DS chamber and SR chamber.

In the DS chamber, Step I, the absorption-released H^+ ions through eqs 2 and 3 were scavenged by the pre-stored OH^- ions. After Step I, HSO_3^- ions predominated in the A-SO_2 solution at pH 5.0~6.0 (27). Clearly, the HSO_3^- ions were electrically balanced by the pre-stored Na^+ ions.

In the DS chamber, Step I, the oxidation-released H^+ ions through eq 4 were scavenged by the OH^- ions that were in situ generated through eq 6, 6', or 7. The A-SO_2 oxidation through eq 4 generated SO_4^{2-} ions as product of Step I. Concurrently, Na^+ ions were released from the salt bridge to balance the SO_4^{2-} ions electrically. In the SR chamber, Step I, accompanied generation of H^+ ions occurred through eq 8. The H^+ ions were balanced electrically by the SO_4^{2-} ions released from the salt bridge (Figure 1).

In the SR chamber, Step II, the H^+ and SO_4^{2-} ions from Step I that were electrically balanced mutually from Step I stayed there. And the mutually balanced Na^+ and SO_4^{2-} ions from Step I were relocated to this chamber. At the same time, H^+ ions continued to be generated through eq 8. Accordingly, SO_4^{2-} ions continued to release from the salt bridge to balance the H^+ ions. Therefore, the electron balance resulted in a mixture of H_2SO_4 and Na_2SO_4 in the SR chamber after Step II. Since a reaction of Na_2SO_4 and H_2SO_4 is often adopted to manufacture $NaHSO_4$ in industry, $NaHSO_4$ might be obtained as a product in the SR chamber after Step II.

Additionally in the DS chamber, Step II, accompanied generation of OH^- ions occurred through eq 6'. The OH^- ions were balanced electrically by the Na^+ released from the salt bridge (see Figure 1).

Process of Electrochemically-driven Conversion of SO_2 to $NaHSO_4$. Up to now, this process design has been chemically substantiated, and the oxidation reaction of A-SO_2 can be optimized. Accordingly, a process of the SO_2 conversion to $NaHSO_4$ was schemed in Figure 5 to show the mass flows of SO_2, OH^- and H^+ ions, while those of Na^+ and SO_4^{2-} ions are presented in Figure 8. Figures 7 and 8 illustrate that the SO_2 gas is converted to $NaHSO_4$ through two stages: (i) SO_2 absorption plus oxidation to Na_2SO_4 in the DS chamber, and (ii) transformation of Na_2SO_4 to $NaHSO_4$ in the SR chamber.

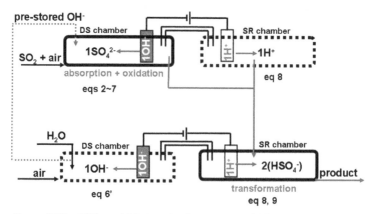

Fig. 7. Mass flow of SO_2, OH^-, and H^+ ions in the process of SO_2 conversion to $NaHSO_4$ in aqueous solution: the number before the species designates their mole mass.

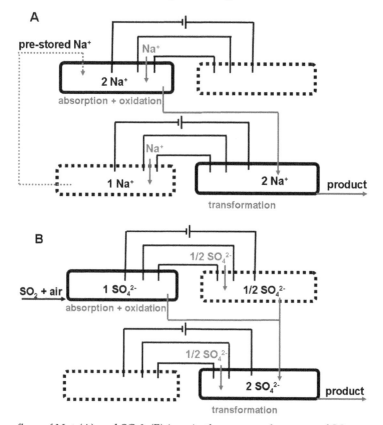

Fig. 8. Mass flow of Na^+ (A) and SO_4^{2-} (B) ions in the proposed process of SO_2 conversion to $NaHSO_4$ in aqueous solution: the number before the species designates their mole mass; chambers in left hand are DS chamber, and in right hand, SR chamber.

Figure 7 illustrates that in the first stage, the SO_2 gas with the co-existing air is introduced into the aqueous solution where the reactions through eqs 2~7 occur to form Na_2SO_4. The SO_2 absorption occurs effectively at high pH, and the A-SO_2 oxidation occurs rapidly at pH 5.0~6.0 but quite slowly at pH 8.0. Thus, although occurring concurrently, the SO_2 absorption predominates first at pH above 6.0, and then the A-SO_2 oxidation becomes the main reaction at pH 5.0~6.0. In the second stage, the solution containing Na_2SO_4 is relocated from the DS chamber to the SR chamber, and then the Na_2SO_4 is transformed into $NaHSO_4$ (eq 9) under an acidic condition through eq 8.

Accompanied with the A-SO_2 oxidation in the DS chamber, H^+ ions are generated (eq 8) in the SR chamber for the subsequent formation of $NaHSO_4$. Accompanied with the $NaHSO_4$ transformation in the SR chamber, OH^- ions are generated (eq 6′) in the DS chamber and pre-stored for the operation of next run.

To practice this new process, some findings from this study may be summarized as tips: (i) alkaline condition at pH > 7.0 in the DS chamber is beneficial to the SO_2 absorption, which can be achieved by the electrochemical pre-storage of OH^- ions through eq 6′, (ii) electrochemical generation of H_2O_2 (eqs 6 and 6′) in the DS chamber can be designed as a concurrent reaction with the A-SO_2 oxidation reaction by air, since the H_2O_2 can accelerate the A-SO_2 oxidation reaction significantly, and (iii) a suitable current density shall be adjusted to maintain the optimal pH 5.0~6.0 of the A-SO_2 oxidation reaction.

Thus by combination of the mass flow in Figure 8, an overall reaction can be expressed below to illustrate the SO_2 conversion into $NaHSO_4$ under the electrical driving force:

$$SO_2 + \frac{1}{2}O_2 + H_2O + Na_2SO_4 \xrightarrow{\quad electricity \quad} 2NaHSO_4 \tag{10}$$

Eq 10 indicates that from a thermodynamic point of view, the electrochemical reactions of eqs 6, 6′, 7 and 8 may not be essential, but from an engineering point of view, they are critical. The cathodic reactions scavenge the absorption- and oxidation-released H^+ ions to drive the SO_2 absorption and oxidation, and the anodic reaction provides H^+ ions to drive the $NaHSO_4$ formation. Therefore, the electrochemical reactions ensure the process of SO_2 conversion to $NaHSO_4$ through eq 10.

Sustainability Evaluation of The Process. Green chemistry has a set of principles to minimize pollution as far as possible in chemical processes, including achieving high value of atom economy in the synthetic process, eliminating toxicity to human health and the environment, minimizing energy consumption, utilizing clean materials, and so on (28,29). Although the aspiration of green chemistry is preferably realized in the manufacturing processes rather than the subsequent cleanup of effluent, the principles are still applicable in the wet FGD process. In this study, sustainability of this new process was evaluated from the perspective of green chemistry.

The atom economy is expressed by AU, which is defined by a ratio of mole mass between the desired product and the reactant(s) as shown below:

$$AU = \frac{mole\ mass\ in\ desired\ product}{mole\ mass\ in\ reac\tan t(s)} \times 100\% \tag{11}$$

The AU values of eqs 1~10 are listed in Table 3. Clearly, eqs 5, 6, 9, and 10 received 100% of AU. Eqs 2, 3, and 4 received AU as high as 98.8%, 98.8%, and 90.0%, respectively. Only eqs 7 and 8 received 50.0% and 11.1%, respectively.

eq	AU (%)	desired product	byproduct
1	75.5	$CaSO_4 \cdot 2H_2O$	CO_2
2	98.8	HSO_3^-	H^+
3	98.8	SO_3^-	H^+
4	90.0	SO_4^{2-}	H^+
5	100	SO_4^{2-}	no
6	100	H_2O_2	no
7	50.0	OH^-	H_2O_2
6'	----	no	H_2
8	11.1	H^+	O_2
9	100	bisulfate	no
10	100	$NaHSO_4$	no

Table 3. Values of Atom Utilization (AU), desired products, and byproducts in eqs 1~10.

In addition to the individual AU values of single reaction, the byproducts of eqs 1~10 are listed in Table 3, and their fates and environmental impacts are discussed as follows. The H^+ ions as byproduct in eqs 2~4 are scavenged to form H_2O, and the H_2O can be reused through eq 8 to generate H^+ ions that entirely end up in the $NaHSO_4$ through eq 9. The H_2O_2 in eq 6 or 6' is reused to enhance the $A\text{-}SO_2$ oxidation and also enters into the $NaHSO_4$. The gaseous byproducts of H_2 in eq 6' and O_2 in eq 8 escape into the environment, whereas it is not hazardous to human health and the environment, while some safety measure should be taken to deal with the H_2. However, all other atoms are kept in the final product through reusing the byproducts in the same setup.

Therefore, it can be seen that except the portion of oxygen and hydrogen atoms that released as gases through eqs 7 and 8, respectively, all other atoms are remained in the final product. This process in essence utilizes the raw materials of Na_2SO_4, O_2 in ambient air, and water which are all environmentally clean. As a result, secondary pollution can be avoided.

The electricity consumption (E) in terms of kWh through eqs 6, 6', 7 and 8 is calculated in light of the two-step model experiment.

In the $A\text{-}SO_2$ oxidation of Step I, electricity was consumed through the cathodic reactions of eqs 6, 6', and 7 to scavenge the H^+ irons released in eqs 2~4, and its consumption was calculated by:

$$E = \frac{\bar{U} \cdot I \cdot t \cdot 10^{-3}}{64 \cdot (C_0 - C_t) \cdot 10^{-3} \cdot S} \tag{S1}$$

where $\bar{U}$ and I were the voltage in average and current that kept constant, respectively, the C_0 and C_t were the $A\text{-}SO_2$ concentrations before and after the oxidation reaction, respectively, and S was the volume.

In this study, $C_0 = 1.0 \times 10^{-3}$ M, $C_t = 0.02\ 10^{-3}$ M, $I = 8.0 \times 10^{-3}$ A, $\bar{U} = 4.0$ V, $t = 0.5$ h, and $S = 0.20$ L; $C_0 = 1.5 \times 10^{-3}$ M, $C_t = 0.18\ 10^{-3}$ M, $I = 10.0 \times 10^{-3}$ A, $\bar{U} = 4.3$ V, $t = 0.5$ h, and $S = 0.20$ L. Thus, in both cases, 1.28 kWh of electricity was consumed in the oxidative conversion of every kg of A-SO_2 to Na_2SO_4 in Step I.

Experimental results showed that the H^+ ions scavenged by eqs 6, 6', and 7 in the A-SO_2 oxidation of Step I were equivalent to the H^+ ions electrochemically generated through eq 8 in Step II. Thus, the E values of the A-SO_2 oxidation and Step II were equal, and doubled electricity of A-SO_2 oxidation, i.e. 2.56 kWh was consumed to convert every of kg A-SO_2 to $NaHSO_4$.

Figure 7 demonstrates that the OH^- ions were pre-stored through eq 6' in Step II. Obviously, the electricity consumption to pre-store the OH^- ions for Step I was included in Step II. Therefore, a total electricity of 2.56 kWh was needed for the overall conversion of every kg of gaseous SO_2 to $NaHSO_4$ in aqueous solution, which was only slightly higher than the electricity consumption between 1.8 and 2.4 kWh for the anodic oxidation conversion of one kg of SO_2 into H_2SO_4 in aqueous solution (14). In comparison, the electricity consumption by the electrochemical reactions of this new process seems quite competitive.

Also important, the final product of $NaHSO_4$ has an added value. Eq 10 indicates that 2.06 kg of Na_2SO_4 needs to be consumed to convert one kg of SO_2 to attain 3.75 kg of $NaHSO_4$ with 82% increase in mass. It is learned from the current market that $NaHSO_4$ has an approximately quadruple commercial value of Na_2SO_4. In addition, the alkaline and acid demanded by this new process are provided on site electrochemically instead of addition of chemicals. Considering that the alkaline and acid are manufactured at the price of electricity, we can presume that this approach appears clean and cheap.

Therefore, this new process can fully comply with the principles of green chemistry and shows promising feasibility. If it is flexibly applied in a wet FGD process for SO_2 removal, it could be an environmentally-sustainable technique. In fact, few of environmental processes, which serve to decompose the pollutants, have high value of atom economy. Fortunately, all the atoms from the raw materials in this process end up in the product of $NaHSO_4$. During the process, some intermediates are involved, while they are reusable and environmentally benign.

Further Investigations. Before this process is practiced for reduction of SO_2 in flue gas, more investigations remain. First, effect of CO_2 which abounds in the flue gas should be precluded. We will demonstrate elsewhere that the CO_2 can be separated advisably from the SO_2 in this process, and then the CO_2 can be further captured and recovered electrochemically in the manner of synchronous supply of alkaline and acid. Second, a side reaction that accompanies the cathodic reaction of O_2 reduction is the H_2 evolution (eq 7). A coupling of this reaction with the anodic oxidation of H_2O (eq 8) shows the well-know reaction of H_2O electrolysis. While eq 7 also outputs OH^-, the reaction of H_2O electrolysis consumes extra energy. Thus, the H_2O electrolysis must be avoided by a careful operation of the electrochemical reactor and selection of cathode on which the H_2 evolution can be inhibited. Third, eq 6 has a byproduct of H_2O_2, which serves to accelerate the S(IV) oxidation. Under some circumstances such as on Pt modified carbon electrode, the O_2 reduction proceeded in a 4-electron pathway to solely output OH^- free of H_2O_2. In this case, the S(IV) oxidation is accomplished by the air oxidation while the overall reaction of $NaHSO_4$ production remains. Since the air oxidation of S(IV) is fast to some extent, and thus in practice, whether the H_2O_2 is essential needs further scrutiny. Forth, a salt bridge is employed in this study to spatially separate the cathodic and anodic chambers. In real application, a membrane *(17,18)* that is commercially available can be employed. Anyway, this new process is promising as an alternative FGD process that immobilizes the SO_2 waste in the form of non-calcium product by means of a cheap and non-toxic material, and thereby avoids the concern over any secondary pollution *(30)*.

5. Nomenclature

A-SO_2 = absorbed SO_2 in aqueous solution
AU = atom utilization in %
CSH = chemical scavenging of H^+ ions through addition of NaOH solution
DS = desulfurization chamber, as cathodic chamber
ESH = electrochemical scavenging of H^+ ions through eqs 6, 6' or 7
FGD = flue gas desulfurization
SH = scavenging of H^+ ions
SR = sulfur-recovery chamber, as anodic chamber

6. Acknowledgements

This work was supported by Natural Science Foundation of China (Project No: 50978260, 21077136).

7. References

[1] Philip, L.; Deshusses, M. A. Sulfur dioxide treatment from flue gases using a biotrickling filter-bioreactor system. *Environ. Sci. Technol.* 2003, *37*, 1978-1982.
[2] Srivastava, R. K.; Jozewicz, W. Flue gas desulfurization: the state of the art. *J. Air Waste Manage. Assoc.* 2001, *51*, 1676-1688.
[3] Lynch, J. A.; Bowersox, V. C.; Grimm, J. W. Acid rain reduced in eastern United States. *Environ. Sci. Technol.* 2000, *34*, 940-949.
[4] Cofala, J.; Amann, M.; Gyarfas, F.; Schoepp, W.; Boudri, J. C.; Hordijk, L.; Kroeze, C.; Li, J.; Lin, D.; Panwar, T. S.; Gupta, S. Cost-effective control of SO_2 emissions in Asia. *J. Environ. Manage.* 2004, *72*, 149-161.
[5] Kikkawa, H.; Nakamoto, T.; Morishita, M.; Yamada, K. New wet FGD process using granular limestone. *Ind. Eng. Chem. Res.* 2002, *41*, 3028-3036.
[6] Gutiérrez Ortiz, F. J.; Vidal, F.; Ollero, P.; Salvador, L.; Cortés, V.; Giménez, A. Pilot-plant technical assessment of wet flue gas desulfurization using limestone. *Ind. Eng. Chem. Res.* 2006, *45*, 1466-1477.
[7] Karatza, D.; Prisciandaro, M.; Lancia, A.; Musmarra, D. Calcium bisulfite oxidation in the flue gas desulfurization process catalyzed by iron and manganese ions. *Ind. Eng. Chem. Res.* 2004, *43*, 4876-4882.
[8] Lancia, A.; Musmarra, D. Calcium bisulfite oxidation rate in the wet limestone-gypsum flue gas desulfurization process. *Environ. Sci. Technol.* 1999, *33*, 1931-1935.

[9] Hrastel, I.; Gerbec, M; Stergaršek, A. Technology optimization of wet flue gas desulfurization process. *Chem. Eng. Technol.* 2007, *30*, 220-233.

[10] Michalski, J. A. Equilibria in limestone-based FGD process: magnesium addition. *Ind. Eng. Chem. Res.* 2006, *45*, 1945-1954.

[11] Rosenberg, H.S. Byproduct gypsum from flue gas desulfurization processes. *Ind. Eng. Chem. Res.* 1986, *25*, 348-355.

[12] Fan, M.; Brown, R. C.; Zhuang, Y.; Cooper, A. T.; Nomura, M. Reaction kinetics for a novel flue gas cleaning technology. *Environ. Sci. Technol.* 2003, *37*, 1404-1407.

[13] Zeman, F. Energy and material balance of CO_2 capture from ambient air. *Environ. Sci. Technol.* 2007, *41*, 7558-7563.

[14] Scott, K.; Taama, W.; Cheng, H. Towards an electrochemical process for recovering sulphur dioxide. *Chem. Eng. J.* 1999, *73*, 101-111.

[15] Struck, B. D.; Junginger, R.; Boltersdorf, D.; Gehrmann, J. The anodic oxidation of sulfur dioxide in the sulfuric acid hybrid cycle. *Intern. J. Hydrogen Energy.* 1980, *5*, 487-497.

[16] Wiesener, K. The electrochemical oxidation of sulphur dioxide at porous catalysed carbon electrodes in sulphuric acid. *Electrochim. Acta* 1973, *18*, 185-189.

[17] Huang, C.; Xu, T.; Yang, X. Regenerating fuel-gas desulfurizing agents by using bipolar membrane electrodialysis (BMED): effect of molecular structure of alkanolamines on the regeneration performance. *Environ. Sci. Technol.* 2007, *41*, 984-989.

[18] Huang, C.; Xu, T. Electrodialysis with bipolar membranes for sustainable development. *Environ. Sci. Technol.* 2006, *40*, 5233-5243.

[19] Connick, R. E.; Zhang, Y. -X.; Lee, S. Y.; Adamic, R.; Chieng, P. Kinetics and mechanism of the oxidation of HSO_3^- by O_2. 1. the uncatalyzed reaction. *Inorg. Chem.* 1995, *34*, 4543-4553.

[20] Liu, H.; Wang, C.; Li, X.; Xuan, X.; Jiang, C.; Cui, H. A novel electro-Fenton process for water treatment: reaction-controlled pH adjustment and performance assessment. *Environ. Sci. Technol.* 2007, *41*, 2937-2942.

[21] Gözmen, B.; Oturan, M. A.; Oturan, N.; Erbatur, O. Indirect electrochemical treatment of bisphenol in water via electrochemically generated Fenton's reagent. *Environ. Sci. Technol.* 2003, *37*, 3716-3723.

[22] Brillas, E.; Calpe, J. C.; Casado, J. Mineralization of 2,4-D by advanced electrochemical oxidation processes. *Water Res.* 2000, *34*, 2253-2262.

[23] Konnick, R. E.; Zhang, Y.-X. Kinetics and mechanism of the oxidation of HSO_3^- by O_2. 2. the manganese (II)-catalyzed reaction. *Inorg. Chem.* 1996, *35*, 4613-4621.

[24] Vidal, B. F.; Ollero, P.; Gutierrez Ortiz, F. J.; Arjona, R. Catalytic oxidation of S(IV) in seawater slurries of activated carbon. *Environ. Sci. Technol.* 2005, *39*, 5031-5036.

[25] Komintarachat, C.; Trakarnpruk, W. Oxidative desulfurization using polyoxometalates. *Ind. Eng. Chem. Res.* 2006, *45*, 1853-1856.

[26] Yu, G.; Lu, S.; Chen, H.; Zhu, Z. Oxidative desulfurization of diesel fuels with hydrogen peroxide in the presence of activated carbon and formic acid. *Energy & Fuels* 2005, *19*, 447-452.

[27] Streeter, I.; Wain, A. J.; Davis, J.; Compton, R. G. Cathodic reduction of bisulfite and sulfur dioxide in aqueous solutions on copper electrodes: an electrochemical ESR study. *J. Phys. Chem. B* 2005, *109*, 18500-18506.

[28] Trost, B. M. The atom economy-a search for synthetic efficiency. *Science* 1991, *254*, 1471-1477.

[29] Lankey, R. L.; Anastas, P. T. Life-cycle approaches for assessing green chemistry technologies. *Ind. Eng. Chem. Res.* 2002; *41*, 4498-4502.

[30] Wang C., Liu H., Li X.Z., Shi J.Y., Ouyang G.F., Peng M., Jiang C.C., Cui H.N., A new concept of desulfurization: the electrochemically driven and green conversion of SO_2 to $NaHSO_4$ in aqueous solution. Environ. Sci. Technol., 2008, 42, 8585-8590.

4

Application of Nanometals Fabricated Using Green Synthesis in Cancer Diagnosis and Therapy

Iliana Medina-Ramirez[1], Maribel Gonzalez-Garcia[2],
Srinath Palakurthi[3] and Jingbo Liu[4,5]
[1]Department of Chemistry,
Universidad Autonoma de Aguascalientes, Aguascalientes,
[2]Department of Chemistry,
Texas A&M University-Kingsville, Kingsville, TX,
[3]Department of Pharmaceutical Sciences,
Texas A&M Health Science Center, Kingssville, TX,
[4]Nanotech and Cleantech Group,
Texas A&M University-Kingsville, TX,
[5]Department of Chemistry,
Texas A&M University, College Station, TX,
[1]Mexico
[2,3,4,5]USA

1. Introduction

The interest for the development of new materials for biomedical applications has steadily increased over the past ten years, due to the numerous advances made in the field of cancer diagnosis and therapy using nanoparticles (NPs). Nowadays, these NPs, such as noble metal gold (Au) and silver (Ag), are considered as valuable starting materials for the construction of innovative nanodevices and nanosystems that are built based on the rational design and precise integration of the tailored-functional properties of NPs. The two main goals of this investigation are to: (1) conduct multidisciplinary project and emerging research in the biological and physical sciences to develop new diagnostic methods or cancer therapy tools (health aspect); and (2) optimize the fabrication variables of nano-metals using green colloidal chemistry method (nanotechnological aspect). To accomplish the above goals, we have extensively investigated the fabrication of nano-structured metal(s) using green colloidal (sol-gel) approaches to formulate particles of specific size with defined homogeneity at molecular level; characterized the fabricated nanostructured materials using state-of-the-art instrumentation; and evaluated their *in vitro* cytotoxicity using model cell lines (such as ovarian adenocarcinoma and normal ovarian cell line), and related hemocompatibility of Au and Ag NPs with human red blood cells. The scope of this article will focus on introductory nanoscience, green synthesis strategies, and structural analysis techniques, followed by specific examples related to diagnostics and cancer therapy.

1.1 Source and properties of engineered nanomaterials

1.1.1 General view of nanotechnology

The development of engineered nanomaterials (ENMs) is considered as one of the major achievements of the twentieth century. [1] The novel and outstanding physicochemical properties (which are distinctively different from that of bulk materials) of these ENMs have led to their use in numerous current and emerging technologies. [2] Nowadays, it is practically impossible to find any field of knowledge that is not in some way or another related with nanomaterials; for instance, development of diagnostic sensors for biomedical and environmental applications. [3] In biology and medicine, sensors are being used as DNA/protein markers for disease identification, or as novel drug carriers with little or no immunogenicity and high cell specificity. [4] In materials science, ENMs are currently being used for the development of solar cells, light-emitting diodes (LEDs), information recording systems and non-linear optical devices. [5] Although ENMs represent numerous advantages in their applications, a number of significant challenges still remain in order to ensure the implementation of synthetic pathways that allow for controlled production of all nanomaterials with desired size, uniform size distribution, morphology, crystallinity, chemical composition, and microstructure, which altogether result in desired physical properties. [6] Another important consideration for the practical application of these materials is the high cost associated to their large scale production, coupled with the tremendous difficulties in separation, recovery, and recycling in industrial applications. [7]

1.1.2 Properties of nanomaterials

Nanostructured materials display three major unique properties not observed in their bulk counterparts. [8] They possess: 1) "ultra high surface effect" allowing for dramatic increase in the number of atoms in the surface. [9] When the nanosize is reduced to about 10 nm, the surface atoms account for 20 % of the total atoms composed of the perfect particles. If the particle size is further decreased to 1 nm, the surface atoms account up to 99 % of the total number of atoms. [10] Due to the lack of adjacent atoms, there exist large amount of dangling bonds, which are not saturated. Those atoms will bind with others to be stabilized. This process results in lower than the maximum coordination number and increased surface energy, collectively resulting in high chemical reactivities of the generated nanomaterials; 2) "ultrahigh volume effect" allowing for light weight of small particles. [11] Due to the reduction in the diameter of the nanomaterials, the energy gap was increased. Herein, the electrons are mobile relative to the bulk. This causes unique physical, chemical, electronic and biological properties of nanomaterials compared with macroscopic systems; and 3) "quantum size effect" allowing for nanosize decrease and quasi-discrete energy of electron orbital around the Femi energy level. [12] This will increase the band gap between highest occupied molecular orbital (HOMO) and lowest unoccupied molecular orbital (LUMO), shown in *Fig. 1*. Therefore, the electromagnetic quantum properties of solids are altered. When the nanometer size range is reached, the quantum size effect will become pronounced, resulting in abnormal optical, acoustic, electronic, magnetic, thermal and dynamic properties. The above energy gap (δ) of conduction and valence band of metals was determined by Kubo using an electronic model, $\delta=4E_f/3N$, where the E_f stands for Fermi enegy, N the total electrons in the particles. [13]

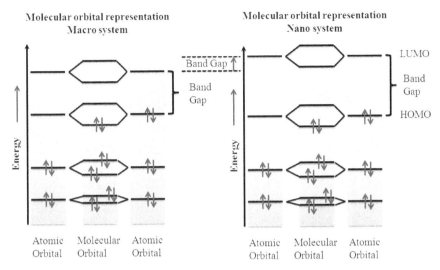

Fig. 1. The band gap between HOMO and LUMO (note: the band gap of nanosystem was increased compared with macrosystem).

1.1.3 Synthesis of nanomaterials

The development of cost-effective environmental friendly methods for large-scale synthesis of benign, highly efficient nanomaterials represents a critical challenge to their practical applications in biomedical research. [14] Some nanoporous materials with regular shapes such as porous nanowires, nanotubes, spheres, and nanoparticles have been successfully prepared by chemical or physical methods which can be carried out in a variety of ways, such as in the gas phase, or in solution, or supported on a substrate, or in a matrix. [15] Although a comprehensive comparison of these approaches does not exist, significant differences in the physicochemical properties, and therefore performance of the resulting materials does exist, allowing for some quantative assessment. In general, physical methods (also known as "top-down" techniques, *Fig. 2*) are highly energy demanding, besides, it is difficult to control the size and composition of the fabricated materials. [16] In the top-down method, the bulk is "broken" down to the nanometer length scale by lithographic or laser ablation-condensation techniques. [17] Chemical methods (also known as "Bottom-up" techniques, *Fig. 2*) are the most popular methods of manufacturing nanomaterials. [18] They are characterized by narrow nanoparticles size distribution, relative simplicity of control over synthesis, and reliable stabilization of nanoparticles in the systems; besides, kinetically controlled mixing of elements using low temperature approaches might yield nanocrystalline phases that are not otherwise accessible. [19] These methods are based on various reduction procedures involving surfactants or templating molecules, as well as thermal decomposition of metal or metal-organic precursors. [20] The sol-gel process has proved to be very effective in the preparation of diverse metal oxide nanomaterials, such as films, particles or monoliths. [21] The sol-gel process consists of the hydrolysis of metal alkoxides and subsequent polycondensation to form the metal oxide gel. [22] One means of achieving shape control is by using a static template to enhance the growth rate of one crystallographic phase over another. [23] The organic surfactants may be undesired for many applications, and a relatively high temperature is needed to decompose the material. [24] Unfortunately, such thermal treatment generally induces dramatic growth of nanoparticles such that ultrafine nanoparticles free of templating and stabilizing agents could not be obtained [25] Lately, novel and simple methods to prepare metal oxides with controllable morphologies by simply varying the hydrolytic conditions have been reported. [26]

Fabrication of Nanoscale Materials

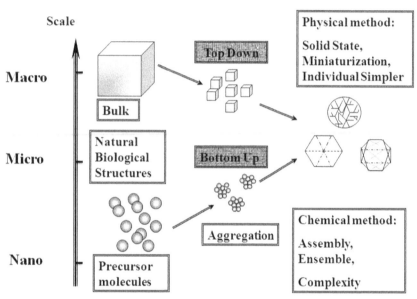

Fig. 2. Schematic of Bottom-up and Top-down Fabrication.

1.2 Toxicological effects of engineered nanomaterials

Metal based nanoparticles (NPs) have been widely used in various applications including biological diagnostics, cell labeling, targeted drug delivery, cancer therapy, and biological sensors and also as antiviral, antibacterial and antifungal agents. [27] An understanding of the potential toxicity induced by these NPs to human health and environment is of prime importance in development of these NPs for biomedical applications. The metal NPs can enter the body via routes such as the gastrointestinal tract, lungs, intravenous injection and exposure in skin. When NPs come to contact with biological membranes, they pose a threat by affecting physiology of the body. For example, silver (Ag), copper (Cu) and aluminum (Al) NPs may induce oxidative stress and generate free radicals that could disrupt the endothelial cell membrane. It was reported recently that in utero exposure to NPs present in exhaust of diesel affects testicular function of the male fetus by inhibiting testosterone production. [28] In this section toxicity issues of each metal nanoparticle will be discussed, starting with aluminum followed by gold, silicon, copper, titania and ceria. Al NPs are widely used in military applications such as fuels, propellants, and coatings. Thus, exposure of aluminum to soldiers and other defense personnel is on the rise. Recent studies by Wagner *et al* showed that Al NPs exhibit higher toxicity in rat alveolar macrophages and also their phagocytic ability is diminished after 24 h of exposure. [29] Al NPs have produced significant increase in lactate dehydrogensae (LDH) leakage and were shown to induce apoptosis after exposing them to mammalian germline stem cells. [30] *In vivo* toxicity experiments of aluminum oxide nanoparticles in imprinting control region (ICR) strained mice indicated that nano-alumina impaired neurobehavioral functions. Furthermore, these defects in neurobehavioral functions were shown to be mediated by mitochondrial impairment, oxidative damage and neural cell loss. [31]

Gold (Au) NPs are recently widely used in cellular imaging and photodynamic therapy. Au NPs exhibited size-dependent toxicity with smaller-sized particles showing more toxicity than larger-sized particles in various cell lines *in vitro*, and *in vivo* similar pattern of size dependent toxicity was observed. [32] The effect of shape of the Au NPs on toxicity was also assessed and it was reported that Au nanorods were more toxic than spherical Au NPs. [33] The effect of surface chemistry of Au NPs was investigated in monkey kidney cells (CV-1) in and cells carrying SV40 genetic material (Cos-1 cells), and *Escherichia coli* (*E. coli*) bacteria. The results indicated that cationic Au NPs were more toxic compared to their anionic counterparts. [34]. Lately, biofuncionalization (Lysine capped) of Au NPs has been explored in an attempt to reduce their toxicity. These lysines capped Au NPs were not toxic to macrophages at concentrations up to 100 μM after 72 h exposure. Moreover, they did not elicit the secretion of pro-inflammatory cytokines tumor necrosis factor-alpha (TNF-α) or interleukin-1 beta (IL-1β). [35]

Silicon microparticles were investigated for their biomedical application; the results have shown that vascular endothelial cells which internalized silicon microparticles maintained their cellular integrity as demonstrated by cellular morphology, viability, and intact mitotic trafficking of vesicles bearing silicon microparticles. [36] Silica (SiO_2) NPs were also investigated for their toxicity as they promise effective biomedical applications. When SiO_2 NPs were treated on normal human mesothelial cells at concentration of 26.7 μg/mL, there was only 3% LDH leakage after 3 hr exposure. When mice were treated with SiO_2-nanoparticle coated magnetic nanoparticles for 4 weeks, the NPs were shown to be taken up by the liver and then redistributed to other organs such as spleen, lungs, heart, and kidney. It was also reported that NPs (<50 nm) bypassed various biological barriers (blood-brain and blood-testis) without inducing any toxicity. [37] The Ag NPs have also shown cellular toxicity; *in vitro* they exhibited size and dose dependent toxicity in neuroendocrine cells, liver cells, lung cells and germline stem cells, and the toxicity was reported to be mediated mainly through oxidative stress. *In vivo*, these NPs of 60 nm size were investigated at various doses (30, 300 and 1000 mg/kg) and these NPs showed a dose dependent liver toxicity. [38] Copper (Cu) nanoparticles are widely investigated for their antimicrobial properties. Cu NPs, though very effective antimicrobials, exhibited severe toxicological effects including heavy injuries in kidney, liver, and spleen of mice after administration. [39,40]

Titanium dioxide (TiO_2) nanoparticles are very widely manufactured nanomaterials for various applications including cosmetics, paints and as additives to surface coatings. TiO_2 NPs have shown to induce inflammatory responses and reactive oxygen species (ROS) in various cell types and tissues. [41] However, Renwick *et al* reported that TiO_2 NPs were not directly toxic to macrophages but significantly reduced the ability of macrophages to phagocytose other particles. [42] Cerium oxide (CeO_2) nanoparticles are recently used in computer chip manufacturing, polishing and as an additive to decrease diesel emissions.

The toxicity studies of CeO_2 nanoparticles involving human lung adenocarcinoma epithelial cell line (A549 lung cells) have shown that CeO_2 NPs did not result in any significant change in LDH leakage or cell morphology, but exhibited a NP induced oxidative stress resulting in altered gene expression. [43]

Hemocompatibility of metal nanoparticles is a very important property for biomedical applications. Metal nanoparticles with good hemocompatibility are often desirable since red blood cells are the primary cells that come in contact with metal nanoparticle when administered intravenously. Various attempts were made to improve the hemocompatibility of metal nanoparticles. It was reported that when Zein, a natural polymer, was incorporated into silver NPs, the hemocompatibility was easily achieved. The results indicated that zein-silver nanocomposites have shown better hemocompatibility when compared with Ag NPs alone. [44,48-49] Ren *et al* have investigated the hemocompatibility of cisplatin loaded Au-Au2S nanoaparticles. The results indicated that bare Au-Au2S NPs have shown hemolysis ratio below 2 % at 100 µg/mL concentration. The cisplatin loaded Au-Au2S nanoparticles have shown hemolysis ratio < 5% at 80 µg/mL, indicating their hemocompatibility and potential use for cancer therapy. [45]

1.3 Nanostructural characterization of engineered nanomaterials

We have been able to synthesize several metallic and semiconducting NPs, which have been evaluated using several state-of-the-art instrumentation techniques. Spectroscopic and microscopic techniques were employed in order to determine the stability, crystalline phase, morphology and size distribution of the synthesized NPs.

1.3.1 Surface energy study of engineered nanomaterials

The electrokinetics (expressed by zetapotential, ζ) of the colloidal suspension was measured using a ZetPALS approach to evaluate particle size and its distribution, from which the stability of engineered nanomaterials can be further determined. [46] Based on the sign of particle's, ζ, the charge can be also determined. [47] The time dependence of the zetapotential on the course of measurement is another technique to understand the formation mechanism of nanoparticles. It was encountered that the large zetapotential of the like (negative) charges enabled to minimize the particles agglomeration due to electrostatic repulsion.

1.3.2 Crystalline phase study of engineered nanomaterials

X-ray powder diffraction (XRD) analysis was used to identify the crystalline phase of the NPs due to its capability to provide rapid, non-destructive analysis of multi-component mixtures, allowing quick and accurate analysis of phase, crystallinity, lattice parameters, expandition tensors and bulk modules, and aperiodical arranged clusters. Based on the peak broadening, the cyrstallite size can also be calculated using Scherrer equation. [50] XRD characterization is widely used in various fields as metallurgy, mineralogy, forensic science, archeology, condensed matter physics, and the biological and pharmaceutical sciences. *Fig. 3* display the working principle and major information obtained from XRD.

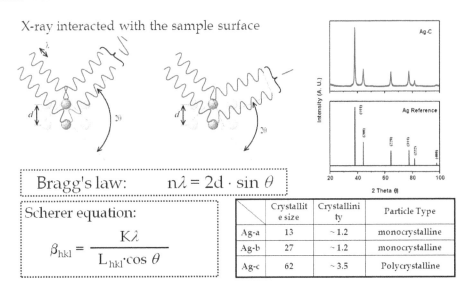

X-ray interacted with the sample surface

Bragg's law: $n\lambda = 2d \cdot \sin\theta$

Scherer equation:

$$\beta_{hkl} = \frac{K\lambda}{L_{hkl}\cdot\cos\theta}$$

	Crystallite size	Crystallinity	Particle Type
Ag-a	13	~1.2	monocrystalline
Ag-b	27	~1.2	monocrystalline
Ag-c	62	~3.5	Polycrystalline

n is an integer determined by the order given, λ is the wavelength of x-rays (nm), and moving electrons, protons and neutrons, d is the spacing between the planes in the atomic lattice (nm), θ is the angle between the incident ray, and the scattering planes (rad) , L is the crystallite size (nm), β is the full width at half maximum (rad), and K is a constant, that varies with the method of taking the breadth (0.89<K<1).

Fig. 3. XRD characterization of nano-materials.

1.3.3 Fine-structure study of engineered nanomaterials

In this study, the morphology and crystalline structure of the engineered nanomaterials were characterized using scanning and transmission electron microscopy (SEM and TEM). [51] Both techniqes are based on the use of a microscope that uses high energy electrons to form an image. Due to their advantages of higher magnification, larger depth of focus, greater resolution, and ease of sample observation, both facilities have been employed widely to determine the crystalline phase, defects, and texture of materials. [52] High resolution field emission TEM was employed to achieve high spatial resolution, high contrast, and unsurpassed versatility. Particularly, the Tecnai F20 G^2 TEM used in this study, which includes a Schottky field emission source, provides ultra-high brightness, low energy spread and very small probe sizes. The Tecnai F20 G^2 used in this study has been designed and preconfigured specifically to meet the strict requirements of nanomaterials. This TEM is also equipped with a robust high-brightness field emission gun, allowing for a wide range of applications, from morphological analysis to defect characterisation. [53]

1.3.4 Elemental composition study of engineered nanomaterials

As complementary techniques, energy dispersive spectroscopy (EDS, equiped with TEM) and X-ray photoelectron spectroscopy were used to accurately determine the elemental composition. [54] The EDS is normally used as a semi-quantitative analysis that allows for determination of the amount and identity of the different elements. The EDS can stimulate the emission of characteristic X-rays from a specimen, herein, a high energy beam of charged particles (electrons, protons) is then produced and focused into the sample. An atom within the sample contains ground state electrons in discrete energy levels or electron shells bound to the nucleus. [55] The incident beam form EDS may excite an electron in an inner shell, ejecting it from the shell and creating an electron hole. An electron from an outer, higher-energy shell then fills the hole. The difference in energy may be released in the form of an X-ray. The number and energy of the X-rays emitted from a specimen can be measured by an energy dispersive spectrometer. As the energy of the X-rays is characteristic of the difference in energy between the two shells, and of the atomic structure of the element from which they were emitted, this allows the elemental composition of the specimen to be measured. The principle of EDS is also shown in *Fig. 4* (see § 1.3.3 *Fine-structure study of engineered nanomaterials*).

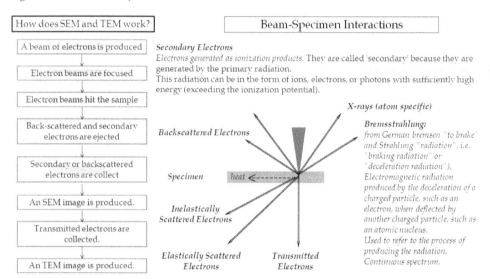

Fig. 4. Schematic demonstrates the working principle of TEM (SEM included).

XPS is a surface chemical analysis technique (*Fig. 5*), which can irradiate a material with a beam of aluminum (Al) or magnesium (Mg) X-rays. [56] Measurement of kinetic energy (KE) and binding energy (BE) can be completed by determining the number of electrons that escape from the top within 1 to 10 nm. Both EDS and XPS require ultra-high vacuum (UHV) conditions to provide uniformity of elemental composition across the top of the surface. [57] Additionally, XPS can also provide uniformity of elemental composition as a function of depth by ion beam ablation and by tilting the sample, empirical formula of pure materials, elements that contaminate a surface, and chemical or electronic state of each element in the surface. An analysis of titanium (Ti) is selected for demonstration. Based on the binding energy, the element can be determined; such as Ti can be identified via measurement of its binding energies of Ti $2p_{3/2}$ and Ti $2p_{1/2}$ electron configurations located at 456.3 electron-volt (eV, ~1.602×10^{-19} C) and 462.2 eV, respectively. In addition, the difference in binding energy can be used for indexing, since the difference in binding energy for the peak splitting of Ti $2p_{3/2}$ and Ti $2p_{1/2}$ was calculated to be 5.9 eV. From this analysis, it can be concluded that Ti was well-indexed with the standard Ti 2p binding energies and their difference ($2p_{3/2}$ = 454.1 eV, $2p_{1/2}$ = 460.3 eV, Δ = 6.2 eV). [58]

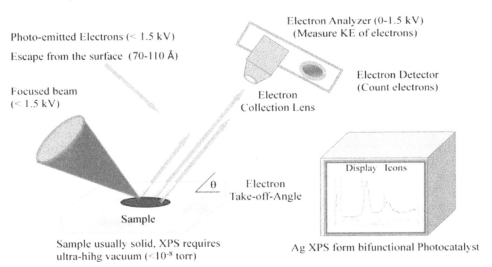

Fig. 5. Demonstration of how XPs works and information obtained.

1.4 Biomedical application of engineered nanomaterials

The most important driving force behind many changes in the field of medical research is the advent of nanotechnology. Progress in nanotechnology is not only aiming at miniaturization but also at systems with increased complexity. This is not just a matter of geometrical structurization but also a matter of specific functionalities that are positioned at discrete locations and in defined distances. Whereas many NMs exhibit localization to diseased tissues via intrinsic targeting, the addition of targeting ligands, such as antibodies, peptides, aptamer, and small molecules, facilitates far more sensitive cancer detection. Nanoparticles with certain specific surface characteristics such as charge and hydrophobicity along with enhanced permeability and retention effect (EPR) have shown higher bioavailability at the target site. EPR is a result of disorganized angiogenesis leading to production of "leaky" blood vessels. EPR effect and lack of effective lymphatic drainage in the tumor tissue have improved the chances of nanoparticle imaging agents with sizes 10-100 nm to be retained in tumor. [59]

Use of nanoparticles in imaging for cancer broadly encompasses two wide areas: (1) detection of certain protein or cancer cells using nanoparticles and, (2) formulation of nano-imaging agents to improve the specificity and to provide high-contrast imaging. Capturing of circulating tumor cells has great potential in cancer diagnosis and therapy. It is very challenging because of the fact that on an average there may be only 1-2 cancer cells per milliliter of blood. Nanotechnology devices based on molecular biomarkers have shown great promise in improving the yield of cancer cell captured. [60] For example, to improve the detection sensitivity and specificity, folate conjugated gold-plated carbon nanotubes were used as contrast agent for photoacoustic imaging. [61]

Many nanoparticle based devices were investigated for their potential applications for imaging in tumor therapy. Most of these were investigated at preclinical stage with few reaching clinical trials. The nanoparticles investigated are either monofunctional or multifunctional. Quantum dots (QDs), which are colloidal fluorescent semiconductor nanocrystals, have sizes ranging from 2-10 nm were investigated in visualization of colon cancer using fiber optics. [62] Iron oxide Nanoparticles conjugated with herceptin as targeting ligand were investigated for detection of small tumors of breast cancer using magnetic resonance imaging (MRI). [63] Dendrimers which are highly branched synthetic polymers were conjugated with prostate specific antigen and were used for imaging in prostate cancer. [64]

Multifunctional nanoparticles are the nanoparticles which combine various functionalities to improve specificity to cancer cells at a single component. Silica based nanoparticles are considered to be promising candidates for development of multifunctional nanoparticles because of their ability to host various materials such as fluorescent dyes, metal ions and drugs. [65] Supramagnetic iron oxide nanoparticles coated with silica were conjugated with fluorescein isothiocyanate to label human bone marrow mesenchymal stem cells. [66]

The most recent nanoparticles also termed as third-generation nanoparticles involve a disease-inspired approach of the "nanocell" which are nanoparticle constructs that comprise a polymeric nanoparticle core enclosed in a lipid-based nanoparticle. For example, an anti-angiogenic agent (combretastatin) will be trapped in the lipid envelope and polymeric nanoparticle core will be loaded with a conventional chemotherapeutic agent like doxorubicin. This nanocell when accumulated in tumor by EPR effect will effectively disrupt the tumor vascular growth by releasing the anti-angiogenic agent followed by release of cytotoxic agent for effective tumor inhibition. [67]

1.4.1 Mechanistic study of cancer theragnostics

Relevant biomedical applications of new nanomaterials are cancer diagnosis and treatment. Nanotechnology offers opportunities to enhance our understanding of the mechanisms of cancer, such as by detecting the generation and distribution of cancer cells in tissues, which can lead to improved diagnosis of this dreadful disease. Furthermore, by harnessing and targeting the toxic properties of nanoparticles, therapeutic agents that are more effective for treating cancer will be developed through molecular targeting.

Cancer encompassess a set of complex and diverse diseases that arise generally when normal cells are transformed into tumorogenic cells, which grow in an uncontrolled fashion to the detriment of the surrounding tissues and eventually the organism. Several factors can play a role in the generation and aggressiveness of the tumor cells, such as accelerated growth, lack of need for growth factors or absence of response to growth inhibitors, unresponsiveness to factors that trigger cell death by apoptosis, enhanced migration properties and ability to invade blood circulation, evasion of the immune response, etc. [68] Mechanistically, the transformed phenotype is derived from the progressive accumulation of genetic mutations, including base changes, nucleotide additions and deletions, insertions, duplications, and chromosomal translocations. [69] Many types of genes have been associated with the development of cancer, such as oncogenes, tumor suppressor genes, apoptosis regulatory genes, cell cycle genes, invasiveness and metastasis related genes, etc. [68-73]

One major difficulty in the treatment of cancer is that the tumor cannot often be detected early enough in patients. Early stage tumors usually have favorable prognosis, while many larger, more developed tumors are frequently refractory to current anti-cancer therapies. Another major problem of many current therapies is the lack of adequate selectivity for cancer cells. This inability to adequately distinguish between cancer and normal, healthy cells, leads to toxic effects on the normal cell population and, therefore, detrimental side effects on the cancer patient. The use of NPs in anti-cancer therapies may improve tumor therapies in these two fronts.

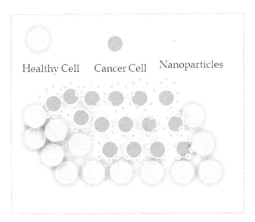

Fig. 6. Detection of cancer cells using NPs.

Different types of metal nanoparticles can be modified to expose molecules that confer them selective binding to specific molecules on the surface of cancer cells, providing probes for the detection and monitoring of cancer cells in tissues (*Fig. 6*). Au NPs in particular have long been utilized in the specific detection of molecules at the surface or inside cells. [74] Other nanomaterials with different chemical reactivity characteristics, including reactivity towards specific functional groups or surface oxidation properties, could provide certain advantages for performing modifications that confer targeting properties to the NPs. The changes in surface plasmon resonance (SPR) properties of Au and Ag NPs caused by interactions with molecules in the surrounding environment can be applied in sensing biomolecules on the surface of tumor cells.

Detectors based on such sensors can be very useful for the early detection of cancer cells (*Fig. 7*). The use of Au NPs is also being explored for the early detection of tumors by other techniques such as X-ray scatter imaging . [75] In this research, hepatocellular carcinoma cells containing Au NPs coated with two layers of charged polymers showed enhanced X-ray scattering over cells containing no gold. Hepatocellular carcinoma is the most common cancer that affects the liver and has an estimated 5-year survival rate of only 10%, since current detection methods can only spot the tumors when they have grown to about 5 centimeters in diameter. The imaging technique utilized by this research group (spatial harmonic imaging, SHI) in combination with targeting of the Au NPs to hepatocellular carcinoma cells via attachment to a specific antibody (FB50) has the potential to detect tumors of only a few millimeters in diameter, which would very significantly improve the prognosis of this cancer type.

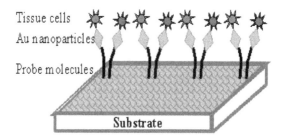

Fig. 7. Schematic design for cancer imaging via optically/electrochemically active gold-NPs.

As indicated, the use of nanomaterials is being investigated not only for the early detection of tumors, but also for the specific destruction of cancerous cells. Ag NPs disrupt the function of proteins and other macromolecules on the surface of bacterial and eukaryotic cells, which can lead to permanent damage and death of the cells. [76-77] Other metallic NPs have different toxicity effects on cells. TiO2 NPs for example do generate reactive oxygen species (ROS) upon absorption of ultraviolet (UV) light. Additionally, specific modifications of the metallic NPs can attach other toxic molecules to the nanomaterial. Combining the toxicity effects of novel nanomaterials with specific targeting to tumor cells can lead to the development of highly specialized therapeutic NPs that can be introduced into the body of cancer patients, where they could detect and destroy cancer cells at early stages, preventing the growth of tumors.

Molecular targeting of NPs was utilized in the generation of TiO2 NPs coupled with an antibody, which targeted the nanomaterial to glioblastoma multiforme cancer cells (glioblastoma multiforme is a type of brain cancer). [78] The antibody used recognizes the interleukin-13 receptor α2 protein chain (IL13Rα2), which has been shown to be almost exclusively overexpressed on the plasma membrane of certain brain tumors, including glioblastoma multiforme. [78-80] The nanomaterial accumulated inside tumor cells and upon exposure to UV light, the NPs led to the formation of ROS and killed the cancerous cells, while they did not show citotoxicity toward normal brain cells.

An example of NPs application for drug delivery is the formation of "Trojan horse" like hybrid NPs. Glycan NPs were synthesized using cyclodextrins as stabilizing agent (polymeric backbone). These glycan NPs were used as envelope for the encapsulation of camptothecin. The formulation of the NPs allowed its internalization on tumor cells; inside the tumor cells, the coating is desintegrated, releasing the toxic molecule that destoys specifically the tumor cell without harming healthy cells. [81] This glycan NPs containing camptothecin, designated as IT-101, have been shown to eliminate or significantly reduce tumors in several mice models of human cancer, and are currently under investigation in clinical trials for the treatment of cancer patients. Delivery of these glycan NPs to tumors is based on the fact that newly formed tumor blood vessels are leaky and let particles as large as 400 to 700 nm into the cancerous mass.

2. Experimental approach

Here, we describe design, synthesis and characterization of nanoscaled metals and their nanostructures. All chemicals used are of analytical grade and were obtained from various vendors. Silver nitrate ($AgNO_3$) and gold (III) chloride trihydrate were purchased from Sigma–Aldrich (St Louis, MO), L-ascorbic acid and other reducing agens from Fisher (Thermo scientific, Pittsburgh, PA), Arabic gum from M.P. Biomedicals, Morgan Irvine, CA), and used without modification. The experiments are carried out for the construction of NPs derived from a bottom-up colloidal method. Non-toxic chemicals and two strong reducants are used in various formulations of NPs, allowing for mitigation of pollution. The structural characterization is by electron microscopy and X-ray diffraction. We also investigated the toxicity of some of the synthesized NPs. Importantly, as previously indicated in the introduction, this approach can also be applied to the toxicity and cancer diagnosis, along the "theragnostics mechanism" described in the following sections.

2.1 Fabrication of nanometals

The experimental workflow to engineer nanoparticles is shown in *Fig. 2* (see §1.1.3 *Synthesis of nanomaterials*). Briefly, the nanometal was created by dissolving different concentrations of silver nitrate ($AgNO_3$) and/or gold (III) chloride trihydrate ($HAuCl_4 \cdot 3H_2O$) in distilled water. Concentration was controlled at 0.005-1.0 M according to application. This aqueous solution was mixed continuously with a magnetic stirrer/mechanical agitation for 30 min. *Arabica* gum (AG, as a surfactant) was added (2 % by mass) to the above solution to improve the dispersion of NPs and to control the particle size. Reducing agents (four selected ones) were then incrementally added into the solution. The molar ratio of metallic cations to reducing agent was controlled at 1:2 to 1:8 and to produce 16 formulations. The colloids were directly used for zetapotential, cytotoxicity and hemocompatibility analyses. The precursor sols were heated between 60 and 80 °C for 2 hrs and filtered and rinsed with ethanol/deionized (DI) water to produce powders used for XRD and and TEM analysis.

2.2 Structural characterization

In tandem to §1.3 (*Nanostructural Characterization of Engineered Nanomaterials*), this section briefly discusses the technical variables or the characterization methods. A ZetaPALS™ (Brookhaven Instruments Corporation, NY) was used to measure the particle size distribution and zeta-potential to evaluate the stability of NPs. The ZetaPALS operating conditions are as follows: An electric field (2 V•cm) was applied to obtain good results. This electric field was used and an alternating current (AC) waveform with the frequency control of 2-20 Hz was applied; measurement duration was kept at 30 to 60 s; laser input was 5 V and laser output power was 35 mW; and an automatic selection of field strength was used. The Ultima III X-ray powder Diffraction (XRD) was employed to determine the crystalline phase and average crystallite size of the NPs. The XRD patterns were collected using a Rigaku multiflex diffractometer, equipped with a copper (Cu) target. The scanning range varied between 30 and 80 degrees at a scanning rate of 2 °/min. The phase structure was then identified through the Jade 7.0 database. A Tecnai F20-G² TEM (FEI Company, Tecnai F20-G² Hillsboro, OR), equipped with ED and EDS techniques were also employed to obtain nanostructural information, crystalline phase and elemental composition of nanometals. In the EDS analysis, the peak intensities were converted to percent weight and then to final molar ratio via the DAX ZAF Quantification tool. The spectra were acquired to determine elemental composition. The HRTEM images were taken at a direct magnification of 600 thousand magnitudes with the point resolution of 0.2 nm. An Axis ULTRA XPS (Kratos

Analytical Inc, NY) was employed to identify the elemental composition of the products. The operating specifications were: high vacuum was controlled under 10^{-8} Torr; anode mode was aluminum (Al, K_α) monochromatic energy source with the power of 10 mA by 12 kV; the lens was used in hybrid mode; resolution of the individual element analysis was of pass energy of 40 eV and 160 eV for survey. [88]

2.3 Evaluation of *in vitro* cytotoxic activity

The cytotoxicity of Au and Ag NPs was performed in two cell lines: ovarian adenocarcinoma cell line (NCI/ADR-RES), and normal ovarian cell line (NCI/CHO). A total of 2×10^4 cells in 200 µL of medium per well were placed in a 96-well plate. After incubation overnight, the medium was replaced with media containing nanometals at different concentrations (0.1, 0.5, 1, 10, and 100 µM) in separate wells. After 24 hrs of incubation, the medium was removed and the cells were washed with ice-cold phosphate-buffered saline (PBS) three times to remove NPs. A volume of 50 µL of 3-[4, 5-dimethylthiazol-2-yl]-2,5,-diphenyl tetrazolium bromide (MTT) at a concentration of 5 mg/mL was then added to each well. Following incubation for 4 hrs, formazan crystals formed were dissolved in 150 µL of dimethylsulfoxide and absorbance was measured at 530 nm using a NOVOstar plate reader (BMG lab technologies, Cary, NC, USA). The percentage viability was calculated by comparing absorbance of treated cells versus untreated control cells which were assigned 100 % viability.

2.4 Evaluation of hemocompatibility

To evaluate hemocompatibility, human red blood cells (HBCs), which composed of reduced leukocytes with adenine-saline were used. HBCs were obtained from coastal bend blood center, Corpus Christi, Texas as a kind gift from. The HBCs were separated from plasma by centrifugation at 4,000 rpm for 5 min at 4 °C. HBCs were washed three times with physiological saline solution and re-suspended in saline to obtain HBCs suspension at 2% (v/v) hematocrit. The so-prepared HBC-suspension was used within 24 hrs upon preparation.

3. Results and discussion

The optimal fabrication variables of NPs will be discussed first (described in § 3.1), followed by electrokinetic properties analysis and crystalline phase strucure of NPs (described in § 3.2-3.3). The *in-vitro* toxicity and hemocompatibility will be discussed next (described in § 3.5) and conclusion on bioapplication of nanomaterials to close the chapter (§ 4.0).

3.1 Fabrication optimization of engineered nanomaterials

The generation process of nanoparticles through facile chemical approach is as follows: (1) A complex between a metal ion and a ligand, that is the functional group of a protecting polymer, is formed in an aqueous solution; (2) The metal ion is reduced to a metal atom with a reducing agent; and (3) The metal atoms aggregate and grow into nanoparticles. The effect of the nature and concentration of the reducing agent was evaluated. As "green" reducing agents, ascorbic acid and sodium citrate were employed. As strong reducing agents, sodium borohydride ($NaBH_4$) and dimethylaminoborane (DMAB) were also tested as the control samples. The concentration of the reducing agent was varied from 1:1, 1:2, 1:4 and to 1:8 molar ratios with respect to metallic ion. Table 1 tabulated the 44 formulations of the nanopartciles via bottom-up "green" colloidal chemistry method. In this study, the reduction of noble metal cation occurs spontaneously. The reactions are shown as follows:

$$Ag^+ (aq) + e^- \rightarrow Ag^\circ (s) \qquad\qquad E^\circ = +0.80V \quad (1)$$

$$AuCl_4^- (aq) + 3\ e^- \rightarrow Au^\circ (s)\ + H^+ (aq) + 4\ Cl^-(aq) \qquad E^\circ = +0.99V \quad (2)$$

$$C_6H_8O_6 (aq) + 2\ e^- \rightarrow C_6H_6O_6 (aq) + 2H^+(aq) \qquad E^\circ = +0.06V \quad (3)$$

According to these three half reaction standard reduction potentials, the overall reaction is determined to have a potential of 0.74 and 0.93 V, respectively. This indicates that the redox reaction between Ag^+ and $C_6H_8O_6$, and $HAuCl_4$ and $C_6H_8O_6$ occurs spontaneously since they are favored, thermodynamically. The released proton H^+ (from both half reactions, the redox process) was responsible for acidity and resulting decrease in pH.

In this research, the dispersing agent, Arabic gum (AG), was used as a size directing agent in the synthesis of the nanometals and to prevent the agglomeration of the fine particles (< 10 nm). Because the complex between surfactant and metal ions can be formed, the growth of the central particle is prevented and terminates at a size in the nanoscale regime (1-100 nm). In short, AG regulates stability of nanoparticles. The synthesis of noble metal nanocomposite via reduction of metal ions in aqueous solutions of AG is based on the formation of a stable metal particle-macromolecular complex.

Cross Product of conc. and molar ratio		Molar ratio of M^{n+} : reducing agent			
		1:1	1:2	1:4	1:8
Concentration of reducing agents and Ag^+ or Au^{3+} (mol/L)	0.005	10 mL * 10 mL**	10 mL 20 mL	10 mL 40 mL	10 mL 80 mL
	0.01	10 mL 10 mL	10 mL 20 mL	10 mL 40 mL	10 mL 80 mL
	0.02	10 mL 10 mL	10 mL 20 mL	10 mL 40 mL	10 mL 80 mL
	0.03	10 mL 10 mL	10 mL 20 mL	10 mL 40 mL	10 mL 80 mL
	. . .	. . .	. . .	. . .	. . .
	0.5	10 mL 10 mL	10 mL 20 mL	10 mL 40 mL	10 mL 80 mL
	1.0	10 mL 10 mL	10 mL 20 mL	10 mL 40 mL	10 mL 80 mL

Note: * represent the volume of metallic cation solution and ** the volume of reducants.

Table 1. Potential set of values of volume of M^{n+}, and reducing agents at designed molar ratio when temperature is maintained at 60 °C and agitation at 1000 rpm.

3.2 Nanocharacterization and bio-application of engineered nanomaterials

The surface energies of nanomaterials are described first in §3.2.1. The crystalline structural, morphological and elemental composition characterization are described next from §3.2.2 to -3.2.4. Lastly, the cytotoxic activity and hemocompatibility of nanometals is described in §3.3 and 3.4, respectively.

3.2.1 Electrokinetic behavior of metallic colloidal suspension

The measured zetapotentials (ζ, with ZetaPALS™) of nanosized metal colloidal suspension were averaged at -44.00 mV. The negative sign indicates large repulsive forces between nanoparticles of either silver or gold preventing flocculation and aggregation of particles and the numerical value indicates samples had colloidal stability (*Fig. 8*). ζ results indicate that the colloid of Ag and Au are stable and agglomeration is successfully prevented using GA as the surfactant. The electrokinetic study also confirms that nano-dispersion of Au with size of 10-15 nm (correspondingly, the particle size of NP powders were found by TEM, see *Fig. 10 a & b*) has been achieved. Fluctuations in the measured value of ζ during the experiment were not observed; therefore the measurements were time independent in the present study, which confirms the stability of nanometal colloid.

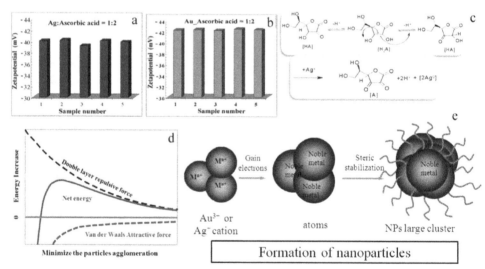

Fig. 8. Relationship between particle size and zetapotential, a: the zetapotential measurement to determine the surface energy of silver nanoparticles, b: the zetapotential measurement to determine the surface energy of gold nanoparticles, c: redox reaction to obtain nanometal particles, d: repulsive and attractive forces in colloidal suspension, and e: schematic of repulsive or attractive forces on nanoparticles, the magnitude which defines aggregation, or separation.

The main goal during the synthesis phase was to control the particle size and its morphology followed by nanocharacterization of the nanoparticles. These particles were synthesized using a bottom-up colloidal chemistry approach. Ag and Au-NP size was controlled through incorporation of GA as the dispersing agent, which also aids wetting of the metal salt. The large negative zetapotential would promote repulsion between the nanoparticles, which in turn would minimize particles agglomeration, which is schematically summarised in *Fig. 8e* (see above figure). The contribution of this green synthesis lies in the following: 1) Finding the best fabrication parameters to achieve complete de-aggregation of the nanoscaled catalyst; 2) Identifying non-toxic dispersing agents to prevent aggregation of the nanocatalyst; and 3) Identifying the suitable reducing agent that ensures complete reduction of metal ions and improves the mono-dispersion of nanocatalyst.

3.2.2 Crystalline phase analyses of engineered nanomaterials

XRD results can provide figure-print characterization of the crystalline phase and the lattice constants. *Fig. 9a* indicates that Ag-NPs display highly crystalline face centered cubic (fcc) space group. The Ag NPs derived from green colloidal chemistry is well indexed with PDF 01-089-3722 (lattice constants, a = 4.0855 Å and α = 90°). In addition, the crystallite size was calculated to be averaged at 30.2 nm. According to the full width at half maximum, the crystallite size was calculated using Scherrer equation (also see Fig 3). Similarly, the Au NPs correspond to the standard PDF 04-0784 (*Fig. 9b*: lattice constants, a = 4.078 Å and α = 90°). The crystallite size of Au nanoparticles was averaged at 10.4 nm; suggesting both Au and

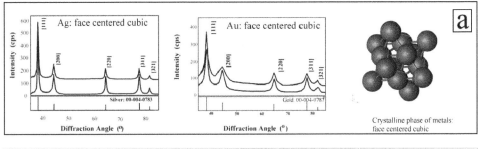

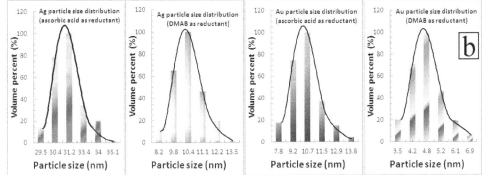

Fig. 9. XRD patterns for nanometals, including XRD patterns for two selected samples, cyrstallite size distribution and cubic centered phase structure, a: XRD pattern of silver and gold nanoparticles and b: particle size distribution of silver and gold nanoparticles.

Ag particle agglomeration was successfully prevented. In both circumstances, the reducing agent, ascorbic acid was selected for demonstration. Other reducing agents provided the same crystalline phase; however, the crystallite size of the engineering particles varies accordingly. It is found that the $NaBH_4$ and DMAB provided essentially the same size, averaged at 12.7 nm for Ag and 4.8 nm for Au nanoparticles, which is also confirmed by TEM measurement. It is important to point out that the lattice occupancy is near ideal (99.0 %), suggesting the Ag and Au nanoparticles possessed highly crystalline structure with less lattice distortion. Both nanoparticles also display high frenquecy of merit (0.2) when using Jade 7.1 database to index with the standard poser diffraction files. The crystalinity index was also determined via taking the ratio of particle size from TEM vs the crystallite size calculated from XRD data. It can be seen that Ag and Au nanoparticles are highly crystalline. Ag showed monoparticles and Au on the other hand polyparticles (see table 2) according to different fabrication methods.

Selected samples	Crystallite size (nm)	Particles size (nm)	Crystallinity	Particle Type
Ag-Ascobic acid	30.2	33.1	≈ 1.1	monocrystalline
Ag- DMAB	12.7	30.4	≈ 2.4	twincrystalline
Ag-DMAB	11.2	34.1	≈ 3.1	polycrystalline
Au-Ascobic acid	10.4	15.6	≈ 1.5	monocrystalline
Au-DMAB	5.8	14.4	≈ 2.2	twincrystalline
Au-DMAB	4.8	15.4	≈ 3.2	polycrystalline

Table 2. Summary of the crystalinity index of Ag and Au nanoparticles (six samples are selected for demonstration)

3.2.3 Fine-structure study of engineered nanomaterials

The morphological and elemental studies through HRTEM technique are shown (Fig.10a and 10b). HRTEM was used as a complementary technique to SEM, which revealed that mono-dispersed and highly crystalline Ag and Au nanoparticles were synthesized. The appearances of both Ag and Au NPs are found to be near-spherical. The particle sizes are varied according to different reducing agents. Generally, the sizes for Ag and Au via reduction of ascorbic acid and citrate are larger than those produced from strong reducing agents, DMAB and $NaBH_4$, with the average particle size from 4.8, to 32 nm in diameter. For both Ag and Ag, ring pattern of the particles indicated that the Ag and Au were well indexed with the standard metallic pattern, which correlated with the XRD analysis.

3.2.4 Elemental composition analyses of engineered nanomaterials

EDS and XPS were used as supplemental techniques to determine the composition of the engineered nanomaterials. EDS spectra of the metal nanoparticles (Fig. 10a & b, lower right-hand-side panel for elemental analysis) indicated the presence of the Ag and Au elements, noting that metallic Ag displays two major emission lines at K_α: 22.162 keV and L_α: 2.984 keV, respectively. While Au displays two major peaks of L_α at 9.711 keV and M_α 2.123 keV, respectively. The characteristic peaks of Au were obtained, and were correlated to K_α peak which occurred at 2.10 KeV and the K_β peak which occurred at 2.18 KeV, respectively, which is in agreement with Au characteristic peaks. The Au NP appearance is near-spherical with the average particle size from 15 to 20 nm in diameter. The high resolution image also shows the lattice fringe which was distinguished, which also confirms the formation of Au crystals. The element Cu peak detected is resulting from Cu grid sample holder.

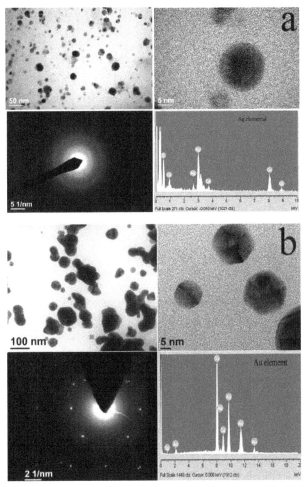

Fig. 10. TEM images and elemental analysis of nanoparticles, a: morphology, ring pattern and composition of nanosilver; b: morphology, ring pattern and composition of nanogold.

The XPS results (*Fig. 11*) show that Ag and Au are characterized by an asymmetric line shape with the peak tailing to the higher BE. The BE may be varied due to its extra columbic interaction between the photoemitted electron and ion core. The binding energies for Ag electron configurations of $3d_{5/2}$ and $3d_{3/2}$ were found to be 366.0 eV and 372.0 eV with the difference of 6.0 eV (Fig. 11a). This measurement was corresponding to the standard Ag 3d binding energies ($3d_{5/2}$ = 368.3 eV, $3d_{3/2}$ = 374.3 eV, Δ = 6.0 eV). Apparently, the 3d photoemission of Ag is split between two peaks with an intensity ratio at about 6:5. It was also found that Au (Fig. 11b) displayed two principal emissions at $4f_{7/2}$ = 82.2 eV, $4f_{5/2}$ = 85.8 eV, respectively. The spectrum was also well-indexed with the standard metallic Au spectra ($4f_{7/2}$ = 82.2 eV, $4f_{5/2}$ = 85.8 eV, Δ = 3.6 eV).

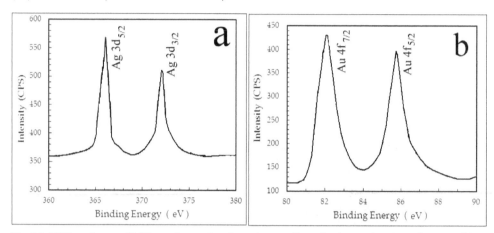

Fig. 11. XPS analyses of NPs, a: determination of nanosilver; b: determination of nanogold.

3.3 *In vitro* cytotoxic activity of nanometals

In addition to nanostructural characterization, the cytotoxicity in normal ovary and ovarian cancer cells using Au and Ag NPs in human red blood cells has been systematically investigated. The *in vitro* cytotoxicity of Ag and Au NPs was assessed in NCI/ADR - RES (OVCAR -8) cells, which are ovarian adenocarcinoma cells, and Chinese hamster ovary (CHO) cells. Samples with 2×10^4 OVCAR-8 and CHO cells were seeded in a 96-well plate. Percentage viability was calculated by comparing absorbance of treated cells with untreated cells (100 % Viability). These results have indicated that at all concentrations (0.1, 0.5, 1, 10 and 100 µM), Au nanoparticles have not shown any toxicity in both normal and ovarian cancer cell lines indicating their biocompatibility. Ag NPs have shown no toxicity at lower concentrations (0.1, 0.5 and 1 µM) in both cell lines. However, at higher concentrations (10 and 100 µM), Ag NPs have shown only 55.9 % and 32.59 % viability in OVCAR-8 cells for 10 and 100 µM doses, respectively. In CHO cells, the Ag NPs have shown 37.88 % and 15.37 % viability for 10 and 100 µM doses, respectively. Based on the data, it may be concluded that the Ag NPs can potentially be used Carriers for cancer therapy. The *in vitro* cytotoxicity data is presented in *Fig. 12a* and *12b*.

3.4 Hemocompatibility of nanometals

Ag and Au NPs were incubated with human red blood cells (RBCs) suspension for 6 hrs in a 96 well plate. The plate was then centrifuged and supernatant was collected and measured for release of hemoglobin by measuring absorbance at 540 nm. RBCs treated with Triton X-100 was used as the reference standard. RBCs are the first point of contact of NPs when administered systemically. In the present study, the Au NPs have shown that the percentage of hemolysis increased with increase in dose. The Au NPs have shown 9.5 % and 12.5 % hemolysis at 0.1 and 1 µM concentrations, respectively. However, at 10 µM concentrations the % hemolysis was found to be 22.4 %. Similarly, Ag NPs have shown hemolytic profile of 9.47 % and 13.07 % at 0.1 and 1 µM. Interestingly, Ag NPs have also shown higher hemolytic activity 29.15 % at 10 µM. These data suggest that at lower concentrations (0.1, 1 µM), Au and Ag NPs can be used for *in vivo* applications. The *in vitro* hemolysis data are presented in *Fig. 13*.

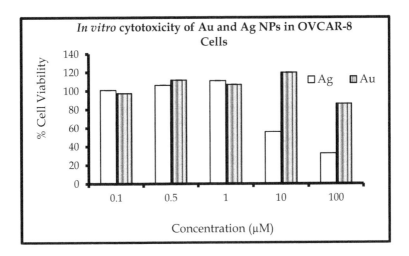

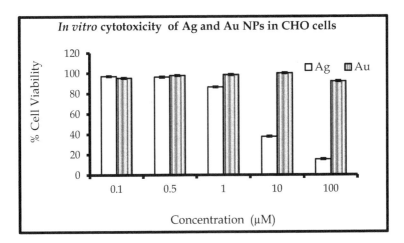

Fig. 12. *In vitro* cytotoxic activity of nanometals, percentage viability was calculated by comparing absorbance of treated cells with untreated cells (100 % Viability), **top panel**: The OVCAR-8 cells were used; **bottom panel**: CHO cells were used.

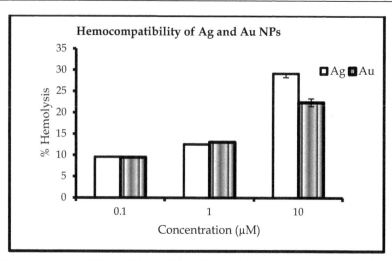

Fig. 13. The hemocompatibility of Ag and Au NPs with human red blood cells, the hollow collum represents the hemocopatibility of Ag, and the filled black colum represents hemocompatibility of Au.

4. Mechanistic studies of nanometals for cancer diagnosis and therapy

A new concept in cancer therapy called "Theragnostics" is emerging recently, which summarizes information based on various biotechnologies involving genomics, transcriptomics, proteomics and metabolomics. [89] The term "Theragnostics" encompasses wide array of topics which includes predictive medicine, personalized medicine, integrated medicine, and pharmacodiagnostics. Major applications of theragnostics in biomedical research include profiling of subgroups of patients based on the likelihood of occurrence of positive outcome to a given treatment so as to entail them to targeted therapies (efficacy), identification of subgroups of patients who are at risk of side effects during a treatment using pharmacogenomics (safety), and monitoring therapeutic response after treatment (efficacy and safety). [90]

The various components of nanotheragnostics each having their own advantage are: signal emitter which emits signal upon excitation by external source, therapeutic moiety can be a drug, or a nucleic acid like small interfering ribonucleic acid (siRNA), nanocarrier is a polymeric carrier capable of carrying high drug payload, and targeting ligand is the entity which can bind to a disease marker with high specificity so as to deliver the entire system to target cell. Non-invasive imaging techniques like magnetic resonance imaging (MRI), X-ray computed tomography (CT), positron emission tomography (PET), single photon emission computed tomography (SPECT), and ultrasound can be used to capture response of signal emitter in real-time. The therapeutic moiety as well as the signal emitter can be encapsulated or covalently attached to the polymeric carrier. Many synthetic and natural polymers were proved to be effective carriers but polymers that are approved for clinical application or currently under clinical trials are poly (ethylene glycol) (PEG), dextran, carboxydextran, β-cyclodextrin, poly (Dextro Levo-lactide-co-glycolide) (PLGA), and poly (L-lysine) (PLL). The most important component of a theragnosite is its targeting ligand. The targeting ligand includes small molecule ligands like short peptides, aptamers, and large molecule ligands like antibodies.

Metallic nanoparticles because of their unique optical properties and their capability to transit therapeutics to tumor have gained wide importance as theragnostic agents. Au nanoparticles possess unique optical properties such as photothermal and surface-plasmon effects, and these properties can be utilized for various clinical applications in cancer diagnostics. Rod shaped Au NPs (Au nanorods) when irradiated in near-infrared region show a surface Plasmon band. This property can be used for bioimaging, or as heating devices for photothermal therapy. Thus, Au nanorods can be used as potential theragnostic devices for bioimaging, thermal therapy and a drug delivery system. When these Au Nanorods coated with as thermosensitive gel were administered to tumor bearing animals and the tumor was irradiated externally, a significantly larger amount of gold was detected in the irradiated tumor. [91] Ag nanosystems have also shown promise as theragnostic agents because of their optical absorption properties. Ag nanosystems were reported to be potential contrast agents for photoacoustic imaging and image-guided therapy. Ag nanosystems consisting of PEGylated Ag layer deposited on silica Cores were investigated and was shown to be non-toxic *in vitro* at concentrations up to 2 mg/mL. Ag nanosystems also have shown strong concentration-dependent photoacoustic signal when embedded in an *ex vivo* tissue sample. [92] Thus, Ag nanosystems which have capability to carry therapeutic payload and exhibit very strong signal when used for image-guided therapy can be widely used for theragnostic applications.

Theragnostics advanced with advent of nanotechnology. Nanotechnology, by virtue of their unique physicochemical properties like quantum confinement in quantum dots, superparamagnetism in certain oxide nanoparticles and surface-enhanced Raman scattering (SERS) in metallic nanoparticles, resulted in emergence of sensitive and cost effective imaging agents. Similarly, their properties like large surface area to volume ratio, capability to control size, hydrophobicity and surface charge according to intended application made them valuable carriers for therapeutic drugs and genes. [93] Many nanocarriers with anticancer drugs are currently investigated clinically for their targeting capability. Immunoliposomes of doxorubicin are in Phase I clinical trials for targeting human metastatic stomach cancer. [94] Polymeric micelles with Paclitaxel are currently in Phase II clinical trials for targeting stomach cancer. [95] Thus, nanoparticles have the required attributes to house therapeutic payload along with diagnostic imaging agent for real-time monitoring of treatment response.

The Au/Ag-NPs exhibit potential for use for molecular-targeted cancer diagnosis and therapeutics. The Au-NPs were biocompatible with the cell lines utilized in our studies. They exhibit strong absorption at 530 nm in the visible light spectrum, with the absorption maximum wavelength displaying shifts upon interaction with other molecules. They are therefore well suited for the development of diagnostic tools aimed at molecular-targeted detection of cancer cells in samples or *in vivo* through imaging technologies. To achieve molecular targeting, modification of the Au-NPs by conjugation with diverse macromolecules (polypeptides, polysaccharides, and even polynucleotides), is being explored in our laboratory. The conjugated Au-NPs can be used in the development of contrast agents for optical imaging to identify the cancerous tissues. On the other hand, Ag-NPs showed cytotoxic effects on both the ovarian adenocarcinoma cell line and the non-tumorogenic ovary cell line in the low micromolar range, as well as stronger hemolytic effects with red blood cells. The toxicity properties of Ag-NPs could be utilized for the development of therapeutic agents to eliminate cancer cells. Once again, conjugation with macromolecules could result in novel nanomaterials with enhanced specificity of action. In this case, the macromolecules should either block the toxicity on normal cells, or target the Ag-NPs to the surface of cancer cells specifically. In the latter case, a lower overall concentration of the targeted Ag-NPs could be utilized, which would reach only toxic levels on the surface or within the cancer cells. Oligosaccharide coatings are currently being investigated in our laboratory. The hypothesis is that cancer cells need to grow fast and would have an enhanced requirement for carbohydrates, driving them to accelerated endocytosis of the nanomaterial, reaching therefore toxic levels of Ag-NPs within the cell upon degradation of the oligosaccharide coat. Furthermore, in advanced cancers the leaky blood vessels would release these NPs right near the cancerous cells, providing for another mechanism for concentration of the therapeutic agent at the target sites.

Our laboratory has focused on the coating of NPs with macromolecules that provide a first level of specificity to distinguish cancer cells from healthy cells. Further specificity for diagnosis or treatment of specific cancer types could be achieved by forming hybrid materials that include specific ligands or antibodies that bind to specific cancer cell types. As described earlier, an antibody against the IL13Rα2 receptor has been utilized to direct TiO$_2$-NPs to certain brain tumors. [78] Our knowledge of molecular markers of specific tumors is increasing steadily, providing tools for the development of a larger collection of high-specificity therapeutic agents that can be utilized in tailored personalized treatment regimens.

5. Conclusion

The green colloidal approach provides stable metallic nanoparticles (Au and Ag). The as-prepared nanometals are determined to be ultrafine, pseudo-spherical and monodispersed. The technical advantages of the colloidal method lie in cost and time effectiveness, simplicity, ability to control homogeneity from molecular level, and high precision of aggregation under use-friendly fabrication variables. The nano-characterization by state-of-the-art instrumentation techniques allow for in-depth understanding of control in morphological, structural and elemental composition via varying the fabrication parameters. Additionally, the Au and Ag nanometals displayed 100 % percentage viability when *in vitro* cytotoxicity was tested. Both metal NPs were also found to be applicable for *in vivo* applications at lower concentrations (0.1, 1 μM).

6. Useful definitions

Phase 0 trials are designed to speed up the development of promising drugs by establishing early on whether they have a similar profile as in the early preclinical studies.

Phase 1 includes trials designed to assess the safety, kinetics, and pharma of a drug.

Phase 2 includes phase I study but on a larger sample size, inculding formulation and dosage.

The Trojan Horse is a stratagem that causes a the nanomaterial to be delivered without detection of the host immune system.

7. Acknowledgements

The authors are grateful to Texas A&M University-Kingsville (TAMUK), the College of Arts and Sciences Research and Development Fund for funding this research activity. The technical support from the National Science Foundation, Major Research Instrumentation program is duly acknowledged to allow the use of JEOL field emission scanning electron microscopy and (TAMUK) Rigaku Ultima III X-ray powder diffraction (TAMU-Corpus Christi). The Academia Mexicana de Ciencias (AMC), Fundación México Estados Unidos para la Ciencia (FUMEC) is also duly acknowledged for their financial support for Dr. Medina-Ramirez to prepare Ag and Au nanoparticles. The Microscope and Imaging Center and the Center of Materials Characterization Facility at TAMU-College Station are also duly acknowledged for technical support and access to the advanced instrumentation, respectively. Last but not least, the authors wish to thank Dr. Sajid Bashir for copy-editing this book chapter.

8. Author contributions

I. Medina-Ramirez conducted the synthesis, and electrokinetic study of nanomaterials. M. Gonzalez-Garcia jointly conceived the conceptual framework with J. Liu and contributed specifically to the application of nanomaterials for cancer diagnosis and treatment. S. Palakurthi completed the toxicicity and hemocompatibility studies. J. Liu prepared Ag and Au nanoparticles based on Medina-Ramirez's discoveries. She completed the spectroscopic and data collection and analysis. She also collected microscopic data with the other investigators (e.g. Dr. Luo at the Materials Characterization Facility and Microscopy Imaging Center) at Texas A&M University-College Station.

9. References

[1] J. Zhang, Z. Liu, B. Han, D. Liu, J. Chen, J. He, T. Jiang, (2004), Chemistry - A European Journal, 10 (14), 3531-3536; Drexler E., Peterson C., Pergamit G. (1991) Unbounding the Future: the Nanotechnology Revolution. William Morrow and Company, New York.

[2] Eric Drexler, Engines of Creation: The Coming Era of Nanotechnology, Bantam Dell Publishing Group Inc (Random House), New York, USA, 1990, pp57-141.

[3] Eric Drexler, Nanotechnology, Molecular Manufacturing, and Productive Nanosystems, John Wiley and Sons Inc, New Jersey, USA, 1991, pp.71-89.

[4] Y. Yao, Y. Ohko, Y. Sekiguchi, A. Fujishima, Y. Kubota, Journal of Biomedical Materials Research Part B: Applied Biomaterials, 2008, 85B, 453.460.

[5] A. Borras, Á. Barranco, J. P. Espinós, J. Cotrino, J. P. Holgado, A. R. González-Elipe, Plasma Processes and Polymers, 2007, 4, 515-527.

[6] F. Zhang, R. Jin, J. Chen, C. Shao, W. Gao, L. Li, N. Guan, Journal of Catalysis, 2005, 232, 424.

[7] A. Kubacka, M. Ferrer, A. Martínez-Arias, M. Fernández-García, Applied Catalysis B: Environmental, 2008, 84, 87-93.

[8] F. Sayılkan, M. Asiltürk, N. Kiraz, E. Burunkaya, E. Arpaç, H. Sayılkan, Journal of Hazardous Materials, 2009, 162, 1309-1316.

[9] A. Mo, J. Liao, W.Xu, S. Xian, Y. Li, S. Bai, Applied Surface Science, 2008, 255, 435.

[10] K. Li, F. S. Zhang, Materials Letters, 2009, 63, 437-441.

[11] F. C. Yang, K. H. Wu, M. J. Liu, W. P. Lin, M. K. Hu, Materials Chemistry and Physics, 2009, 113, 474.

[12] M. Kim, J. W. Byun, D. S. Shin, Y. S Lee, Materials Research Bulletin, 2009, 44, 334-338.

[13] N. Sharifi, N. Taghavinia, Materials Chemistry and Physics, 2009, 113, 63-66.

[14] H. Sakai, T. Kanda, H. Shibata, T. Ohkubo, M. Abe, Journal of American Chemical Society, 2006, 128, 4944-4945.

[15] M. Anpo, M. Takeuchi, Journal of Catalysis, 2003, 216, 505-516.

[16] Y. Ao, J. Xu, D. Fu, C. Yuan, Journal of Physics and Chemistry of Solids, 2008, 69, 2660-2664.

[17] D. Fang, K. Huang, S. Liu, Z. Li, Journal of Alloys and Compounds, 2008, 464, L5-l9.

[18] H. Shibata, T. Ohkubo, H. Kohno, Journal of Photochemistry and Photobiology A: Chemistry, 2006, 181, 357-362.

[19] L. Ge, M. Xu, H. Fang, Journal Sol-Gel Sci Techn, 2006, 40, 65.

[20] D. Wang, F. Zhou, C. Wang, W. Liu, Microporous and Mesoporous Materials, 2008, 116, 658-664.

[21] N. Sobana, K. Selvam, M. Swaminathan, Separation and Purification Technology, 2008, 62, 648.

[22] N. Seirafianpour, S. Badilescu, Y. Djaoued, R. Brüning, S. Balaji, M. Kahrizi, V. Truong, Thin Solid Films, 2008, 516, 6359-6364.

[23] H. Cheng, K. Scott, Journal of Applied Electrochemistry, 2006, 36, 1361-1366.

[24] S. Krishnan, A. C. Kshirsagar, R. S. Singhal, Carbohydrate Polymers, 2005, 62, 309-315.

[25] G. Zhao, S.E. Stevens Jr., BioMetals, 1998, 11, 27-32.

[26] X. G Hou, M. D. Huang, X. L. Wu, A. D. Liu, Chemical Engineering Journal, 2009, 146, 42.48.

[27] J. Sá, C. A. Agüera, S. Gross, J. A. Anderson, Applied Catalysis B: Environmental, 2009, 85, 192-200.

[28] P. Giannouli, R. K. Richardson, E. R. Morris, Carbohydrate Polymers, 2004, 55, 367-377.

[29] M. J. López-Muñoz, J. Aguado, R. V. Grieken, J. Marugán, Applied Catalysis B: Environmental, 2009, 86, 53-62.

[30] S. Bassaid, D. Robert, M. Chaib, Applied Catalysis B: Environmental, 2009, 86, 93-97.

[31] S. K. Ritter, Chemical and Engineering News, 2001, 79(29), 27-34.

[32] L. Picton, I. Bataille, G. Muller, Carbohydrate Polymers, 2000, 2, 23-31.

[33] I. Medina-Ramirez, S. Bashir, Z. Luo, J. Liu, Colloids and Surfaces B: Biointerfaces, 2009, 73, 185-191.

[34] Schrand AM, Rahman MF, Hussain SM, et al. Metal-based nanoparticles and their toxicity assessment.Wiley Interdiscip Rev Nanomed Nanobiotechnol 2010;2:544-68.

[35] Li C, Taneda S, Taya K, *et al* .Effects of in utero exposure to nanoparticle-rich diesel exhaust on testicular function in immature male rats.Toxicol Lett 2009;185:1-8.

[36] Wagner AJ, Bleckmann CA, Murdock RC, *et al*. Cellular interaction of different forms of aluminum nanoparticles in rat alveolar macrophages.J Phys Chem B 2007;111:7353-9.

[37] Braydich-Stolle L, Hussain S, Schlager JJ, *et al*. *In vitro* cytotoxicity of nanoparticles in mammalian germline stem cells.Toxicol Sci 2005;88:412-9.

[38] Zhang QL, Li MQ, Ji JW, *et al*. *In vivo* toxicity of nano-alumina on mice neurobehavioral profiles and the potential mechanisms.Int J Immunopathol Pharmacol 2011;24:23S-29S.

[39] Pan Y, Neuss S, Leifert A, *et al*. Size-dependent cytotoxicity of gold nanoparticles.Small 2007;3:1941-9.

[40] Wang S LW, Tovmachenko O, Rai US, Yu H, Ray PC. Challenge in understanding size and shape dependent toxicity of gold nanomaterials in human skin keratinocytes.Chemical Physics Letters 2008;463:145-49.

[41] Goodman CM, McCusker CD, Yilmaz T, *et al*. Toxicity of gold nanoparticles functionalized with cationic and anionic side chains.Bioconjug Chem 2004;15:897-900.

[42] Shukla R, Bansal V, Chaudhary M, *et al*.. Biocompatibility of gold nanoparticles and their endocytotic fate inside the cellular compartment: a microscopic overview.Langmuir 2005;21:10644-54.

[43] Serda RE, Ferrati S, Godin B, *et al*. Mitotic trafficking of silicon microparticles.Nanoscale 2009;1:250-9.

[44] Kim JS, Yoon TJ, Yu KN, *et al*. Toxicity and tissue distribution of magnetic nanoparticles in mice.Toxicol Sci 2006;89:338-47.

[45] Kim YS, Kim JS, Cho HS, *et al*. Twenty-eight-day oral toxicity, genotoxicity, and gender-related tissue distribution of silver nanoparticles in Sprague-Dawley rats. Inhal Toxicol 2008;20:575-83.

[46] Chen Z, Meng H, Xing G, *et al*. Acute toxicological effects of copper nanoparticles *in vivo*.Toxicol Lett 2006;163:109-20.

[47] Meng H, Chen Z, Xing G, *et al*. Ultrahigh reactivity provokes nanotoxicity: explanation of oral toxicity of nano-copper particles.Toxicol Lett 2007;175:102-10.

[48] Renwick LC, Donaldson K and Clouter A. Impairment of alveolar macrophage phagocytosis by ultrafine particles.Toxicol Appl Pharmacol 2001;172:119-27.

[49] Rothen-Rutishauser B, Grass RN, Blank F, *et al*. Direct combination of nanoparticle fabrication and exposure to lung cell cultures in a closed setup as a method to simulate accidental nanoparticle exposure of humans.Environ Sci Technol 2009;43:2634-40.

[50] Zhang B, Luo Y and Wang Q.Development of silver-zein composites as a promising antimicrobial agent.Biomacromolecules 2010;11:2366-75.

[51] Ren L, Huang XL, Zhang B, *et al*. Cisplatin-loaded Au-Au2S nanoparticles for potential cancer therapy: cytotoxicity, *in vitro* carcinogenicity, and cellular uptake.J Biomed Mater Res A 2008;85:787-96.

[52] E. Traversa, M. L. Vona, P. Nunziante, S. Licoccia, Journal of Sol-Gel Science and Technology, 2001, 22, 115-123.

[53] A. Fujishima, K. Honda, Nature, 1972, 238, 37-38.

[54] V.A. Nadtochenko, A.G. Rincon, S.E. Stanca, J. Kiwi, Journal of Photochemistry and Photobiology A: Chemistry, 2005, 169, 131-137.

[55] T. Saito, T. Iwase, I. Horie, T. Morioka, Journal of Photochemistry and Photobiology B: Biology, 1992. 14, 369-379.

[56] H. Wang, J. Niu, X. Long, Y. He, Ultrasonics Sonochemistry, 2008, 15, 386-392.

[57] G. M. Dougherty, K. A. Rose, J. B. H. Tok, S. S. Pannu, F. Y. S. Chuang, M. Y. Sha, G. Chakarova, S. G. Penn, Electrophoresis, 2008, 29, 1131-1139.

[58] Teoh, W. Yang , Flame spray synthesis of catalyst nanoparticles for photocatalytic mineralization of organics; Ph.D. Thesis, University of New South Wales, 2007, Chapter 1.

[59] J. M Wu, H. C Shih, W.T Wu, Nanotechnology 2005, 17, 105–109.

[60] E.Thimsen, S. Biswas, C. S. Lo, P. Biswas, Journal of Physical Chemistry. C ,2009, 113, 2014–2021.

[61] P.Sangpour, F. Hashemi, A. Z. Moshfegh, Journal of Physical Chemistry. C ,2010, 114, 13955–13961.

[62] S. Shena; L. Guo; X. Chen; F.Ren; C. X. Kronawitter; S.S. Mao, International Journal of Green Nanotechnology: Materials Science & Engineering, 2009, 1, M94 − M104.

[63] X-G Hou, A-D Liu, M-D Huang , B Liao, X-L Wu, Chinese Physics Letters, 2009, 26, 077106-1 to 077106-4.

[64] B.Liang, S. Mianxin, Z. Tianliang, Z.Xiaoyong, D. Qingqing, Journal of Rare Earths, 2009, 27, 461 -468.

[65] S. Chrétien,H.Metiu, 2006, Catalysis Letters, 107, 143-147.

[66] O. Akhavan, Journal of Colloid and Interface Science, 2009, 336, 117-124.

[67] M. S. Chun, H. I. Cho, I. K. Song, Desalination, 2002, 148, 363-367.

[68] Hanahan D, Weinberg RA, 2000. The hallmarks of cancer. Cell 100, 57-70

[69] Vogelstein B, Kinzler KW, 1993. The multistep nature of cancer. Trends Genet 9, 138-141

[70] Vaux DL, Cory S, Adams JM, 1988. Bcl-2 gene promotes haemopoietic cell survival and cooperates with c-myc to immortalize pre-B cells. Nature 335, 440-442

[71] Green DR, Evan GI, 2002. A matter of life and death. Cancer Cell 1, 19-30

[72] Danial NN, Korsmeyer SJ, 2004. Cell death: critical control points. Cell 116, 205-219

[73] Adams JM, Cory S, 2007. The Bcl-2 apoptotic switch in cancer development and therapy. Oncogene 26, 1324-1337

[74] Hermann R, Walther P, Muller M, 1996. Immunogold labeling in scanning electron microscopy. Histochem Cell Biol 106, 31-39

[75] Rand D, Ortiz V, Liu Y, Derdak Z, Wands JR, Taticek M, Rose-Petruck C, 2011. Nanomaterials for X-ray Imaging: Gold Nanoparticle Enhancement of X-ray Scatter Imaging of Hepatocellular Carcinoma. Nano Lett 11, 2678-2683

[76] Bashir S, Chamakura K, Perez-Ballestero R, Luo Z, Liu J, 2011. Mechanism of Silver Nanoparticles as a Disinfectant. International Journal of Green Nanotechnology 3, 118-133

[77] Chamakura K, Perez-Ballestero R, Luo Z, Bashir S, Liu J, 2011. Comparison of bactericidal activities of silver nanoparticles with common chemical disinfectants. Colloids Surf B Biointerfaces 84, 88-96

[78] Rozhkova EA, Ulasov I, Lai B, Dimitrijevic NM, Lesniak MS, Rajh T, 2009. A high-performance nanobio photocatalyst for targeted brain cancer therapy. Nano Lett 9, 3337-3342

[79] Debinski W, Gibo DM, 2000. Molecular expression analysis of restrictive receptor for interleukin 13, a brain tumor-associated cancer/testis antigen. Mol Med 6, 440-449

[80] Mintz A, Gibo DM, Slagle-Webb B, Christensen ND, Debinski W, 2002. IL-13Ralpha2 is a glioma-restricted receptor for interleukin-13. Neoplasia 4, 388-399

[81] Davis ME, 2009. Design and development of IT-101, a cyclodextrin-containing polymer conjugate of camptothecin. Adv Drug Deliv Rev 61, 1189-1192

[82] Lee JH, Huh YM, Jun YW, et al.Artificially engineered magnetic nanoparticles for ultra-sensitive molecular imaging.Nat Med 2007;13:95-9.

[83] Thomas TP, Patri AK, Myc A, et al. In vitro targeting of synthesized antibody-conjugated dendrimer nanoparticles.Biomacromolecules 2004;5:2269-74.

[84] Tallury P, Payton K and Santra S.Silica-based multimodal/multifunctional nanoparticles for bioimaging and biosensing applications.Nanomedicine (Lond) 2008;3:579-92.

[85] Lu CW, Hung Y, Hsiao JK, et al. Bifunctional magnetic silica nanoparticles for highly efficient human stem cell labeling.Nano Lett 2007;7:149-54.

[86] Sengupta S, Eavarone D, Capila I, et al. Temporal targeting of tumour cells and neovasculature with a nanoscale delivery system.Nature 2005;436:568-72.

[87] Ozdemir V, Williams-Jones B, Glatt SJ, et al. Shifting emphasis from pharmacogenomics to theragnostics.Nat Biotechnol 2006;24:942-6.

[88] Pene F, Courtine E, Cariou A, et al.Toward theragnostics.Crit Care Med 2009;37:S50-8

[89] Niidome T, Shiotani A, Akiyama Y, et al.[Theragnostic approaches using gold nanorods and near infrared light].Yakugaku Zasshi 2010;130:1671-7.

[90] Homan K, Shah J, Gomez S, et al. Silver nanosystems for photoacoustic imaging and image-guided therapy.J Biomed Opt 2010;15:021316.

[91] Ho D, Sun X and Sun S. Monodisperse Magnetic Nanoparticles for Theranostic Applications.Acc Chem Res 2011.

[92] Matsumura Y, Gotoh M, Muro K, et al. Phase I and pharmacokinetic study of MCC-465, a doxorubicin (DXR) encapsulated in PEG immunoliposome, in patients with metastatic stomach cancer.Ann Oncol 2004;15:517-25.

[93] Matsumura Y. Poly (amino acid) micelle nanocarriers in preclinical and clinical studies.Adv Drug Deliv Rev 2008;60:899-914.

[94] Holzwarth Uwe and Gibson Neil. The Scherrer equation versus the 'Debye-Scherrer equation' Nature Nanotech 2011; 6:534.

[95] Csáki A, Möller R, Straube W, Köhler JM, and Fritzschea W (2001) DNA monolayer on gold substrates characterized by nanoparticle labeling and scanning force microscopy. Nucleic Acids Res. 16:1-5; Garzon´ IL, Artacho E, Beltran´MR, Garc´ia A, Junquera J, Michaelian K, Ordejon´P, Rovira C, Sanchez-Portal´ D, Soler JM (2001) Hybrid DNA-gold nanostructured materials: an ab initio approach. Nanotechnology 12:126–131.

[96] Chen CS (2008) Biotechnology: Remote control of living cells. Nature Nanotechnology 3:13 – 14; Li Z, Jin R, Mirkin CA, Letsinger RL (2002) Multiple thiol-anchor capped DNA–gold nanoparticle conjugates. Nucleic Acids Research 30:1558-1562.

5

Recent Advances in the Ultrasound-Assisted Synthesis of Azoles

Lucas Pizzuti[1], Márcia S.F. Franco[1], Alex F.C. Flores[2],
Frank H. Quina[3] and Claudio M.P. Pereira[4]
[1]*Universidade Federal da Grande Dourados, Mato Grosso do Sul*
[2]*Universidade Federal de Santa Maria, Rio Grande do Sul*
[3]*Instituto de Química, Universidade de São Paulo, São Paulo*
[4]*Universidade Federal de Pelotas, Rio Grande do Sul*
Brazil

1. Introduction

The ever increasing awareness of the need to protect natural resources through the development of environmentally sustainable processes and the optimization of energy consumption has guided the actions of both the private and governmental sectors of society. Economic planning has been strongly impacted by this new paradigm, which has led to increasing demands by society for products produced in a sustainable way and to more stringent governmental regulatory policies. Thus, while in the past profit was often the major concern, in the current economic context more sustainable production processes are preferred. This has triggered a demand in both industry and academy for the development of new, cleaner technologies.

In the field of chemistry and chemical technology, the 12 principles of Green Chemistry provide a set of clear guidelines for the development of new synthetic methodologies and chemical processes and for the evaluation of their potential for environmental impact. As a consequence, in organic chemistry, numerous investigations now routinely use non-tradicional synthetic metodologies such as solvent-free reactions, the application of alternative activation techniques like microwaves or ultrasound, the replacement of volatile organic solvents by water, ionic liquids, or supercritical CO_2, etc.

In medicinal chemistry, experience has shown that compounds with biological activity are often based on heterocyclic structures. In particular, azoles and their derivatives have attracted increasing interest as versatile intermediates for the synthesis of biologically active compounds such as potent antitumour, antibacterial, antifungal, antiviral and antioxidizing agents. Azoles are a large class of 5-membered ring heterocyclic compounds containing at least one nitrogen atom and one heteroatom in their structure. The construction of this type of molecule has received great attention due to the wide spectrum of biological activities that have been attributed to structurally distinct azoles. Fluconazole, itraconazole, voriconazole and posaconazole are antifungal agents commercially available that contain a triazole nucleus. Celecoxib is a non-steroidal anti-inflammatory and analgesic agent of the pyrazole class. Isoxazole compounds such as valdecoxib are selective COX-2 inhibitors used in the treatment of pain. In this context, much attention has been given by researchers in universities and pharmaceutical industries to the development of new, energy saving, cost-effective, environmentally safe technologies for the synthesis of azoles.

In this context, the use of ultrasound to accelerate reactions has proven to be a particularly important tool for meeting the Green Chemistry goals of minimization of waste and reduction of energy requirements (Cintas & Luche, 1999). Applications of ultrasonic irradiation are playing an increasing role in chemical processes, especially in cases where classical methods require drastic conditions or prolonged reaction times. The excellent review of Cella and Stefani (Cella & Stefani, 2009), which covered the available literature up to about three years ago, clearly showed the importance of taking advantage of the unique features of ultrasound-assisted reactions to synthesize heterocyclic ring systems. The present chapter will therefore limit its coverage of the literature to the period of the last three years and focus on the use of ultrasound to promote the cyclization reactions employed to obtain azoles. Modifications of side chains are not covered in this work. After a brief consideration of ultrasound and the origin of its effects on chemical reactions, a total of 42 reports of the preparation of azoles under conditions of ultrasonic irradiation are reviewed, including several of our own contributions to this field of research. These reports were grouped together according to the number of heteroatoms present in the ring (2, 3 or 4) and each group subdivided by the azole class. Reports in which more than one class of azoles were prepared were collected in the last section, labelled "Miscellaneous".

2. Ultrasound and its chemical effects

The discovery of the piezoelectric effect in the 1880s provided the basis for the construction of modern ultrasonic devices. Piezoelectric materials generate mechanical vibrations in response to an applied alternating electrical potential. If the potential is applied at sufficiently high frequency, ultrasonic waves are generated. The phenomenon responsible for the beneficial effects of ultrasound on chemical reactions is cavitation. Ultrasonic waves are propagated *via* alternating compressions and rarefactions induced in the transmitting medium through which they pass. During the rarefaction cycle of the sound wave, the molecules of the liquid are separated, generating bubbles that subsequently collapse in the compression cycle. These rapid and violent implosions generate short-lived regions with local temperatures of roughly 5000 °C, pressures of about 1000 atm and heating and cooling rates that can exceed 10 billion °C per second. Such localized hot spots can be thought of as micro reactors in which the mechanical energy of sound is transformed into a useful chemical form. In addition to the generation of such hot spots, there can also be mechanical effects produced as a result of the violent collapse (Mason & Lorimer, 2002).

More than 80 years have passed since the effect of ultrasound on reaction rates was first reported by Richards and Loomis (Richards & Loomis, 1927). However, this work received little attention at the time because it used a high-frequency apparatus that was not commonly available to chemists. According to the review of Cravotto and Cintas (Cravotto & Cintas, 2006), two classical papers, published in 1978 and 1980, provided a major stimulus for the development of modern sonochemistry: (1) the report by Fry and Herr (Fry & Herr, 1978) of the reductive dehalogenation of dibromoketones with mercury dispersed by ultrasound; and (2) the work of Luche and Damiano (Luche & Damiano, 1980) on the sonochemical preparation of organolithium and Grignard reagents and their coupling with carbonyls. In the ensuing four decades, numerous efficient and innovative applications of ultrasound in organic synthesis have appeared, which have established sonochemistry as an important tool in the arsenal of Green Chemistry.

The two main sources of ultrasound in organic synthesis are ultrasonic cleaning baths and ultrasonic immersion probes, which typically operate at frequencies of 40 and 20 kHz, respectively (Mason, 1997). The former are more commonly employed in organic synthesis simply because they are less expensive and hence more widely available to chemists, even though the amount of energy transferred to the reaction medium is lower than that of ultrasonic probe systems, which deposit the acoustic energy directly into the reaction medium.

3. Azoles with two heteroatoms

3.1 Pyrazole derivatives

1,5-Diarylpyrazoles (3) can be prepared by the reaction between Baylis-Hillman adducts (1) and phenylhydrazine hydrochloride (2) in 1,2-dichloroetane under sonication with reaction times of 60-180 minutes (Scheme 1) (Mamaghani & Dastmard, 2009). The reactions proceeded regioselectively to afford the desired products in 80-90% yields. The same reaction carried out by simply heating the reaction mixture (80 °C) produced the products in lower yields (60-75%) and required longer reaction times (6-9 h).

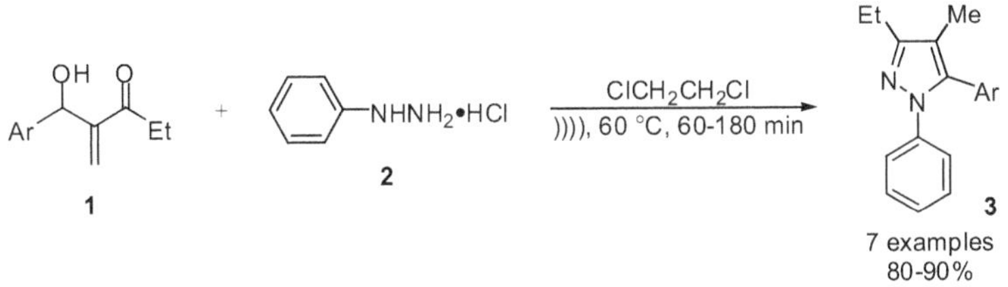

7 examples
80-90%

Scheme 1.

In 2009, Pathak and co-workers (Pathak et al., 2009) conducted a comparative study between four activating methods for obtaining N-acetyl-pyrazolines (7), including reflux, solvent-free conditions, microwave irradiation and ultrasonic irradiation. Microwave irradiation was found to be the most efficient activating method, followed by ultrasound. Employing ultrasound, the reactions of 1,4-pentadien-3-ones (4) with hydrazine (5) and acetic acid (6) in ethanol went to completion in 10-25 minutes and afforded the products (7) in good yields (Scheme 2).

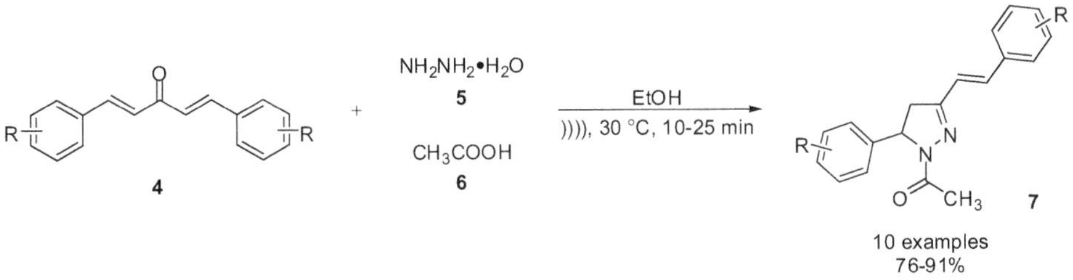

10 examples
76-91%

Scheme 2.

Previously, we described a greener, ultrasound-assisted synthesis of 1-thiocarbamoyl-3,5-diaryl-4,5-dihydropyrazoles (10) from chalcones (8) and thiosemicarbazide (9) catalyzed by KOH (Scheme 3) (Pizzuti et al., 2009). The products were obtained in high purity and in good yields in only 20 minutes *via* a simple filtration of the reaction mixture.

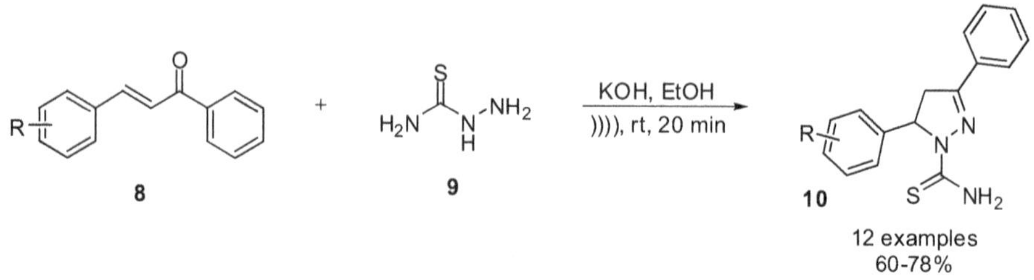

Scheme 3.

Similarly, in 2010, we reported the cyclization of chalcones (11) with aminoguanidine hydrochloride (12) under essentially the same conditions (Scheme 4) (Pizzuti et al., 2010). The 4,5-dihydropyrazole derivatives (13) were obtained in high yields in 30 minutes employing sonication. The same reactions carried out under reflux without ultrasonic irradiation afforded the products in lower yields (57-69%) and required substantially longer reaction times (3-6 h).

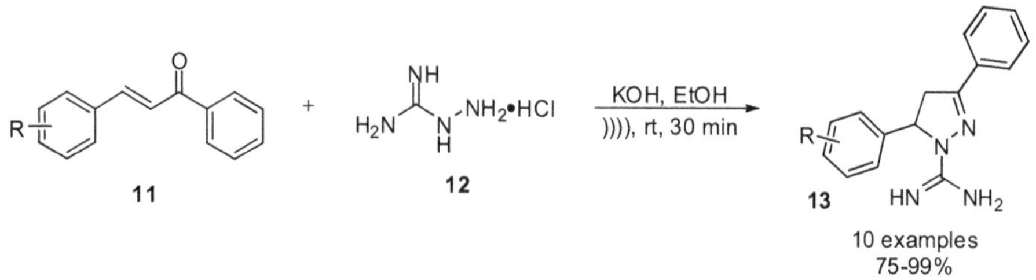

Scheme 4.

Gupta and co-workers utilized an ultrasonic cleaning-bath to promote the cyclization reaction between chalcones (14) and phenylhydrazine (15) under acid conditions, giving the desired cyclization products (16) in good yields (Scheme 5) (Gupta et al., 2010).

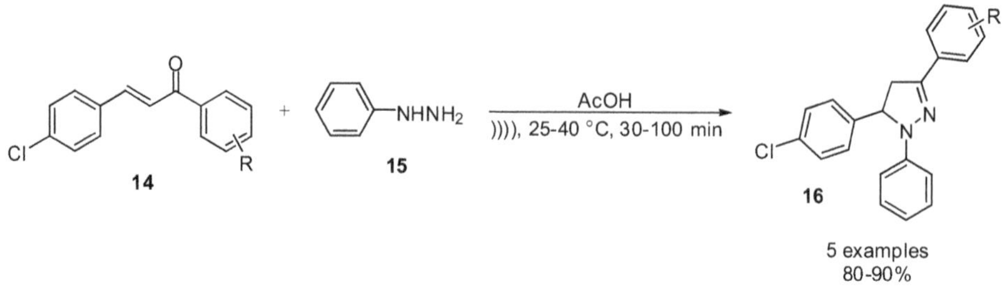

Scheme 5.

A multicomponent ultrasound-assisted protocol for the synthesis of bridgehead pyrazole derivatives (**20**) was developed by Nabid and co-workers (Nabid et al., 2010). Under ultrasonic irradiation, the reaction between phthalhydrazide (**17**), malononitrile or ethyl cyanoacetate (**18**), and aromatic aldehydes (**19**) in the presence of triethylamine furnished 1*H*-pyrazolo[1,2-*b*]phthalazine-5,10-diones (**20**) in very good yields (Scheme 6).

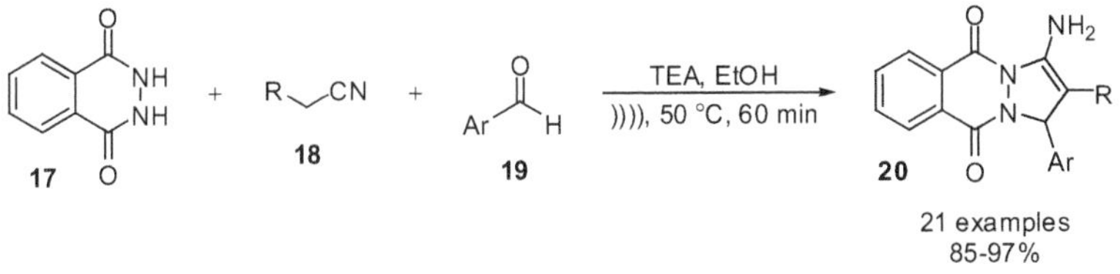

Scheme 6.

3-Aryl-2,3-epoxy-1-phenyl-1-propanone (**21**) reacted under ultrasonic irradiation with phenylhydrazine (**22**) catalyzed by HCl at room temperature to produce 1,3,5-triarylpyrazoles (**23**) in 69-99% yields (Scheme 7) (Li et al., 2010). The same reactions performed in the absence of sonication gave substantially poorer yields.

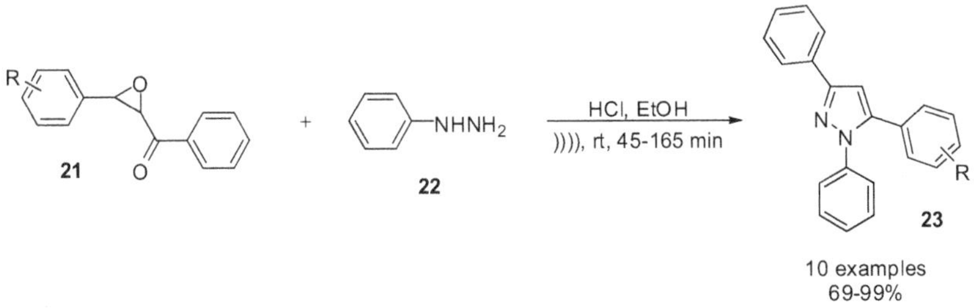

Scheme 7.

3,4-Dimethyl-2,4-dihydropyrazolo[4,3-*c*][1,2]benzothiazine 5,5-dioxide (**26**) was prepared in only 10 minutes by the cyclization of 1-(4-hydroxy-2-methyl-1,1-dioxido-2*H*-1,2-benzothiazin-3-yl)ethanone (**24**) with hydrazine (**25**) under ultrasonic irradiation (Scheme 8) (Ahmad et al., 2010). The product was used for the preparation of acetohydrazide derivatives with potential antioxidant and antibacterial activities.

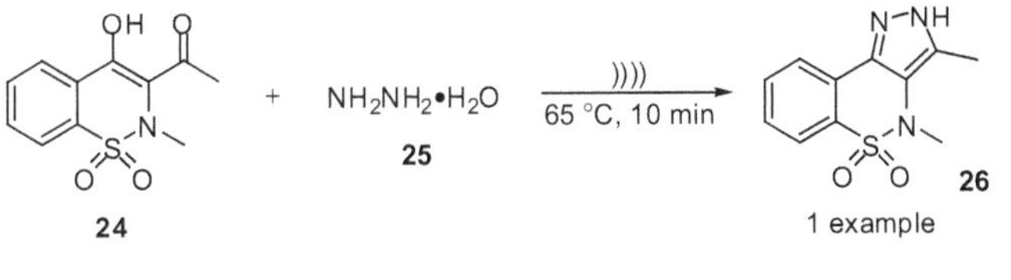

Scheme 8.

In 2010, ultrasound was employed to promote the synthesis of pyrazolones (**29**) *via* the reaction of β-keto esters (**27**) with hydrazine derivatives (**28**) in ethanol. The reactions went to completion in short times (2-25 min) and afforded the products in 4-93% yields (Scheme 9) (Al-Mutairi et al., 2010).

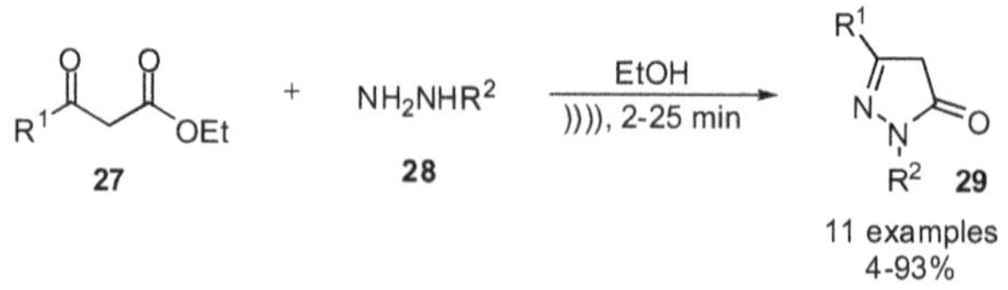

Scheme 9.

Recently, Machado and co-workers (Machado et al., 2011) reported the preparation of several ethyl 1-(2,4-dichlorophenyl)-1*H*-pyrazole-3-carboxylates (**32**) that are structurally analogous to the CB1 receptor antagonists used in the treatment of obesity. Cyclization of 4-alkoxy-2-oxo-3-butenoic ester (**30**) and 2,4-dicholorophenyl hydrazine hydrochloride (**31**) under sonication (10-12 min) or conventional thermal conditions (2.5-3 h) regioselectivelly afforded the desired products (Scheme 10). The use of ultrasound proved to be fundamental for reducing the reaction time.

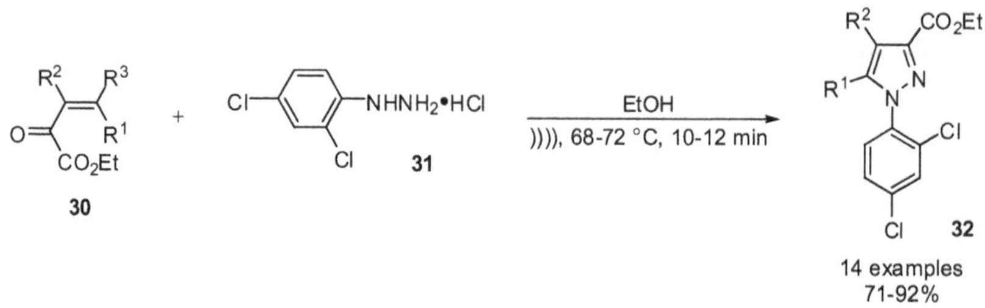

Scheme 10.

A rapid procedure for obtaining acetylated *bis*-pyrazole derivatives (**36**) was based on the sonication of *bis*-chalcones (**33**) and hydrazine (**34**) in the presence of acetic anhydride (**35**) during 10-20 minutes (Scheme 11) (Kanagarajan et al., 2011). The same reactions required 5-8 h to go to completion when carried out under heating in the absence of ultrasound and afforded lower yields (55-70%) than those of sonochemical-assisted reaction.

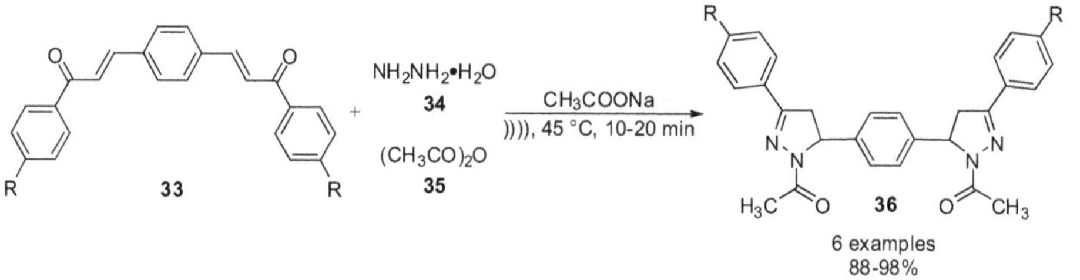

Scheme 11.

A four-component one-pot reaction of ethyl acetoacetate (**37**), aromatic aldehydes (**38**), hydrazine (**39**), and malononitrile (**40**) in water afforded dihydropyrano[2,3-*c*]pyrazoles (**41**) in good yields under ultrasonic irradiation (79-95%) (Scheme 12) (Zou et al., 2011). Again, a comparative study in the absence of ultrasound showed that the products were obtained in lower yields (70-86%) and demanded longer reaction times (1-5 h).

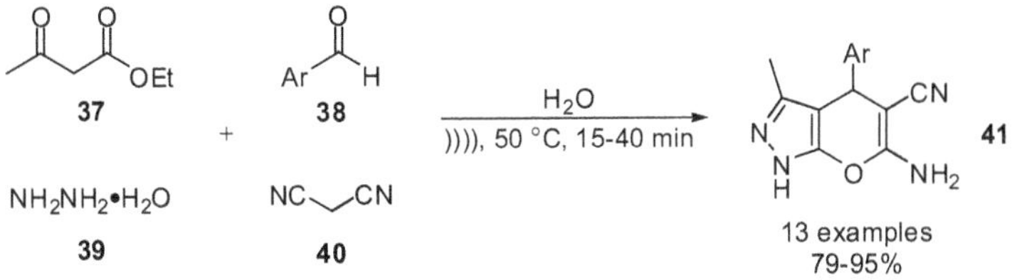

Scheme 12.

Both microwave and ultrasonic irradiation promoted de reaction of nitrile derivatives (**42**) with hydrazines (**43**) (Rodrigues-Santos & Echevarria, 2011). The reaction was highly regioselective and produced only one isomer (**44**) in 70-95% yields after 3 hours under sonication (Scheme 13). Although microwave irradiation required a much shorter reaction time (15 min), the yields were much lower (40-65%) than with ultrasound. However, ultrasound was not effective for the preparation of phenylhydrazine derivatives.

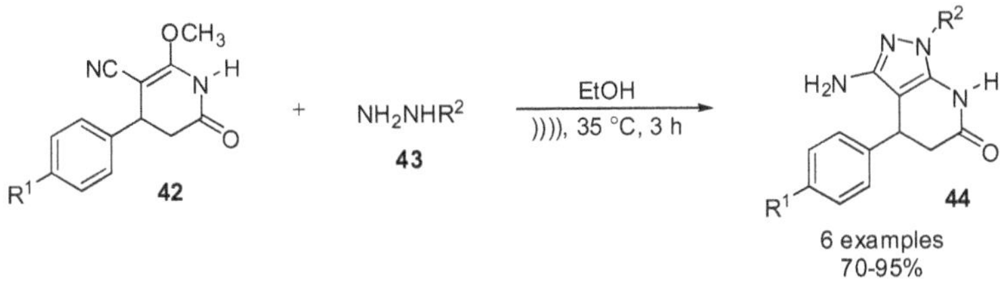

Scheme 13.

Very recently, Shekouhy and Hasaninejad (Shekouhy & Hasaninejad, 2012) reported the rapid and efficient preparation of 2H-indazolo[2,1-b]phthalazine-triones (**49**) (Scheme 14). Their four-component one-pot methodology consisted of the reaction of phthalic anhydride (**45**), dimedone (**46**), and hydrazine hydrate (**47**) with several aromatic aldehydes (**48**) in an ionic liquid under sonication. The products were obtained in good to excellent yields in very short reaction times.

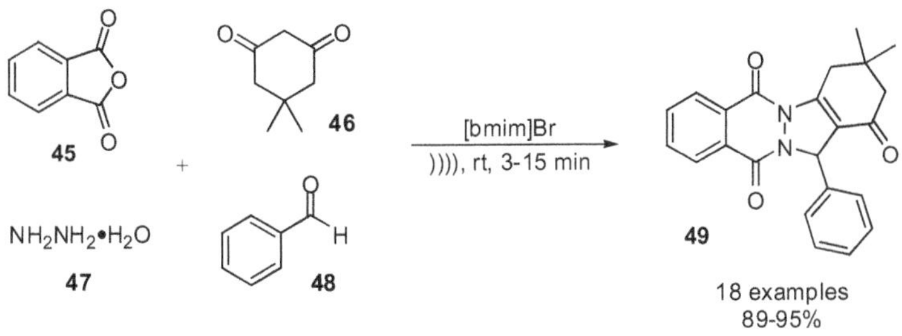

Scheme 14.

3.2 Imidazole derivatives

In connection with reactions in aqueous media, low potency (50 Watts) sonochemistry has been used to prepare 2-imidazolines (**52**) from the reaction of aldehydes (**50**) with ethylenediamine (**51**) and NBS (*N*-bromosuccinimide) as catalyst (Scheme 15) (Sant'Anna et al., 2009). The compounds were isolated in high yields (80-99%) and required only short reaction times (12-18 minutes). The isolated compounds showed bioactivity as monoamine oxidase (MAO) inhibitors.

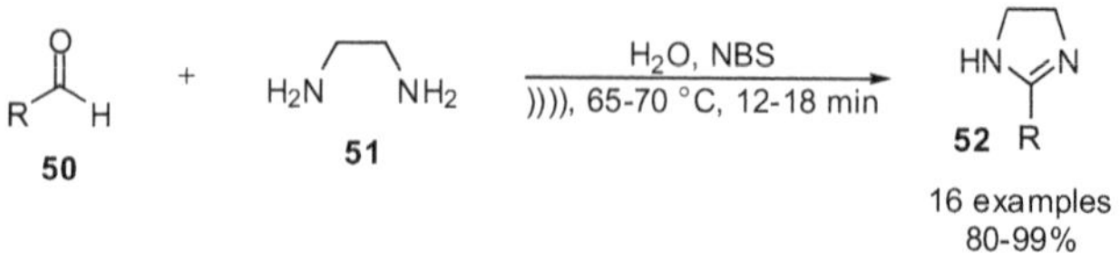

Scheme 15.

Shelke and co-workers (Shelke et al., 2009) reported the synthesis of 2,4,5-triaryl-1*H*-imidazoles (**57**) from the three-component one-pot condensation of benzil (**53**)/benzoin (**54**), aldehydes (**55**) and ammonium acetate (**56**) in aqueous media under ultrasound at room temperature (Scheme 16). BO$_3$H$_3$ (5 mol%) was used as catalyst. The reaction, performed under conventional stirring without ultrasound, required a reaction time (180 minutes) clearly longer than those require when ultrasound was used.

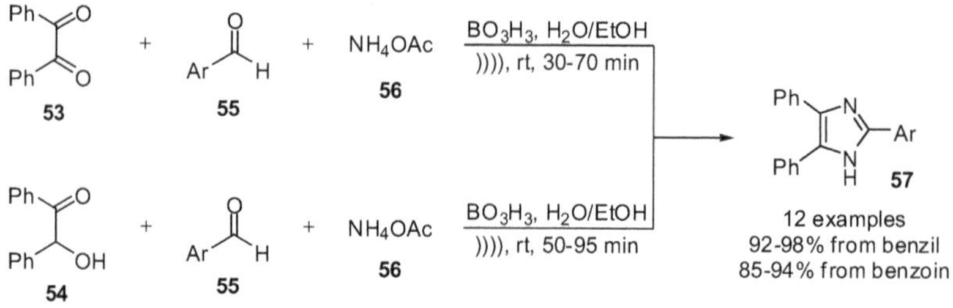

Scheme 16.

Li and co-workers (Li et al., 2010) reported the synthesis of glycoluril derivatives catalyzed by potassium hydroxide in EtOH under ultrasonic irradiation (Scheme 17). Although the reaction was relatively efficient, it was not selective. Two products were isolated, the desired glycoluril (**60**) (17-75%) together with a hydantoin co-product (**61**) (1-37%).

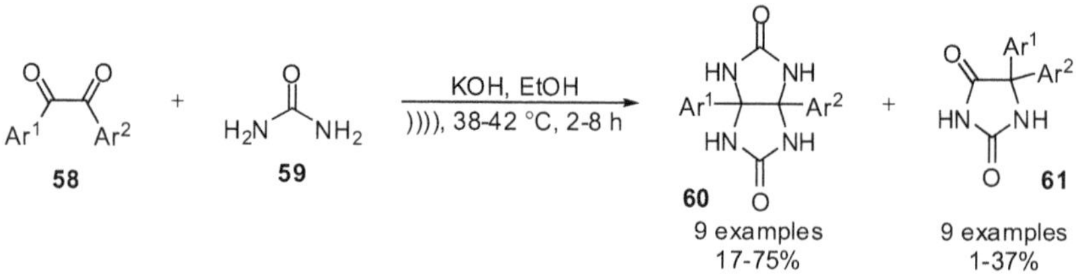

Scheme 17.

Zang and co-workers (Zang et al., 2010) reported a three-component one-pot synthesis of 2-aryl-4,5-diphenyl imidazoles (**65**) at room temperature under ultrasonic irradiation (Scheme 18). The ionic liquid 1-ethyl-3-methylimidazole acetate ([emim]OAc) was used as catalyst and the desired products obtained in satisfactory yields. Ionic liquids have shown great potential as catalysts and are a particularly attractive alternative to conventional catalysts. Their ability to dissolve a wide variety of substance and their potential for recyclability are among the attributes responsible for their recent popularity.

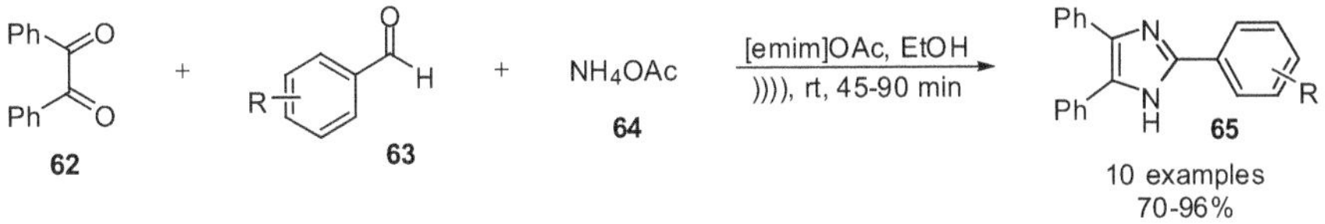

Scheme 18.

Joshi and co-workers (Joshi et al., 2010) reported the synthesis of 1,3-imidazoles (**68**) by the reaction of substituted aldehydes (**67**) with o-phenylenediamine (**66**) catalysed by 5 mol% of tetrabutylammonium fluoride (TBAF) in water under ultrasonic irradiation at room temperature (Scheme 19). Quaternary ammonium fluoride salts are inexpensive and relatively non-toxic reagents and water is generally recognized to be a green solvent in organic synthesis. The reported synthesis thus represents a mild, chemoselective method for preparing these heterocycles.

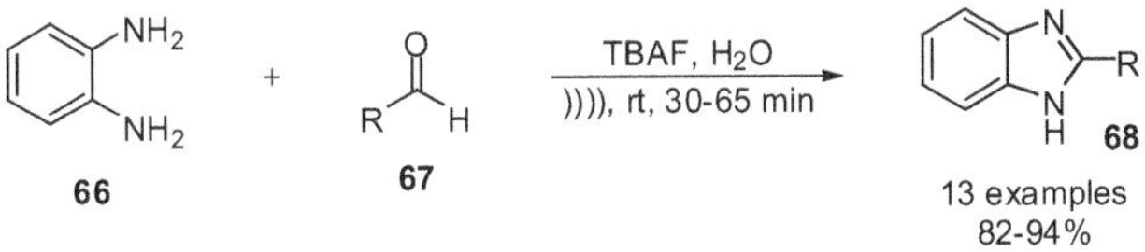

Scheme 19.

Very recently, Arani and Safari (Arani & Safari, 2011) reported a very efficient high yield (96-98%) synthesis of 5,5-diphenylhydantoin (**72**) and 5,5-diphenyl-2-thiohydantoin (**73**) derivatives under mild conditions (Scheme 20). The reactions were performed in DMSO/H$_2$O with ultrasonic irradiation and catalyzed by KOH.

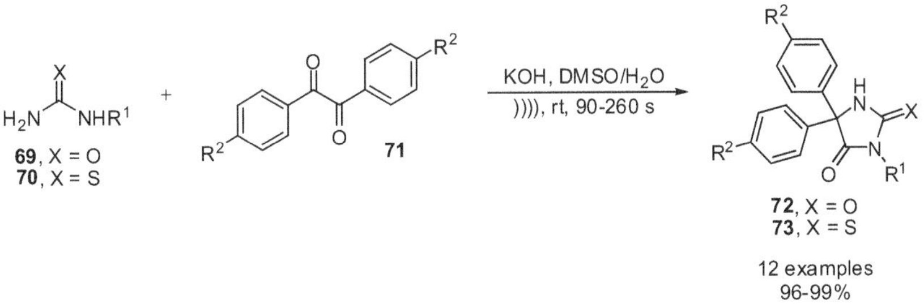

Scheme 20.

3.3 Isoxazole derivatives

Recently, Li and co-workers described a facile and economical procedure for the synthesis of spiro azole compounds (**77**) (Li et al., 2010). The one-pot synthesis of 3-aza-6,10-diaryl-2-oxa-spiro[4.5]decane-1,4,8-trione (**77**) from 1,5-diaryl-1,4-pentadien-3-one (**74**), dimethyl malonate (**75**), and hydroxylamine hydrochloride (**76**) in the presence of sodium hydroxide gave good yields at 50 °C under ultrasound irradiation (Scheme 21).

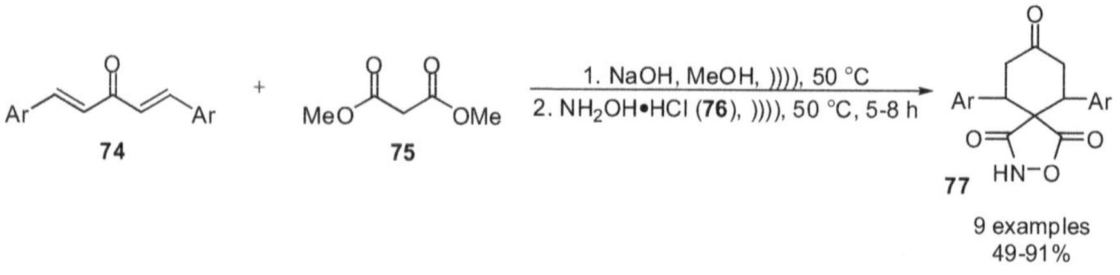

Scheme 21.

A series of dihydroisoxazole derivatives (**80**) were prepared by the ultrasound-promoted cyclization reaction between chalcones (**78**) bearing a quinolinyl substituent and hydroxylamine hydrochloride (**79**) in the presence of sodium acetate in aqueous acetic acid solution (Scheme 22) (Tiwari et al., 2011). The sonochemical method gave better yields (87-90%) of the products in shorter times (90-120 min) than the corresponding thermal reactions (72-78% in 6-7 h).

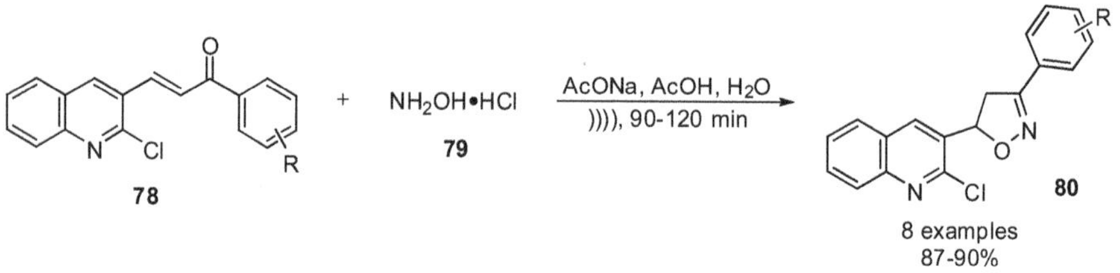

Scheme 22.

3.4 Oxazole derivatives

Ultrasound proved to be efficient for accelerating the cyclization reaction of aryl and methyl nitriles (**81**) with ethanolamine (**82**) catalyzed by $InCl_3$ to give oxazole derivatives (**83**) (Scheme 23) (Moghadam et al., 2009). Products were obtained in 81-97% yields after 5-45 minutes under sonication at room temperature.

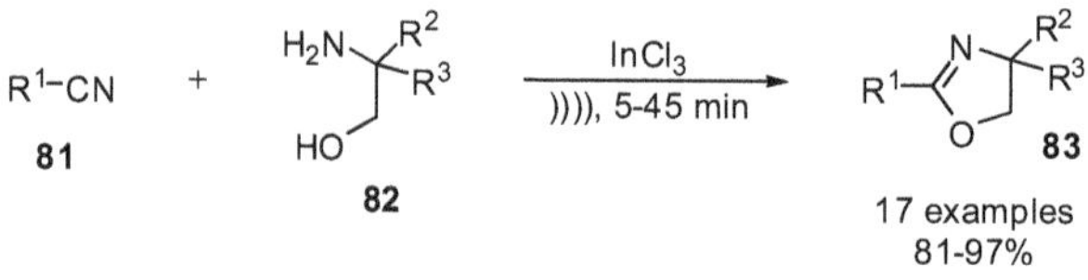

Scheme 23.

The same research group developed a new highly sulfonated carbon-based solid acid and employed it to catalyze reactions similar to those presented above under ultrasonic irradiation (Scheme 24) (Mirkhani et al., 2009). Reactions performed with a combination of the new catalyst and sonication were more efficient than those run without ultrasonic irradiation.

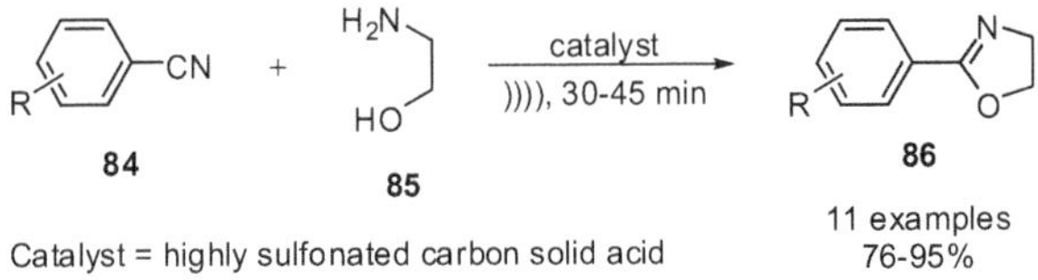
CN
+
H_2N
HO
catalyst
)))), 30-45 min
R
N
O
84
85
86
11 examples
76-95%
Catalyst = highly sulfonated carbon solid acid

Scheme 24.

3.5 Thiazole derivatives

In 2009, Noei and Khosropour (Noei & Khosropour, 2009) reported a high yield, green protocol for the synthesis of 2,4-diarylthiazole derivatives (**89** and **91**) *via* the reaction of arylthioamides (**88** and **90**) with *a*-bromoacetophenones (**87**) under ultrasonic irradiation in the ionic liquid $[bmim]BF_4$ (Scheme 25).

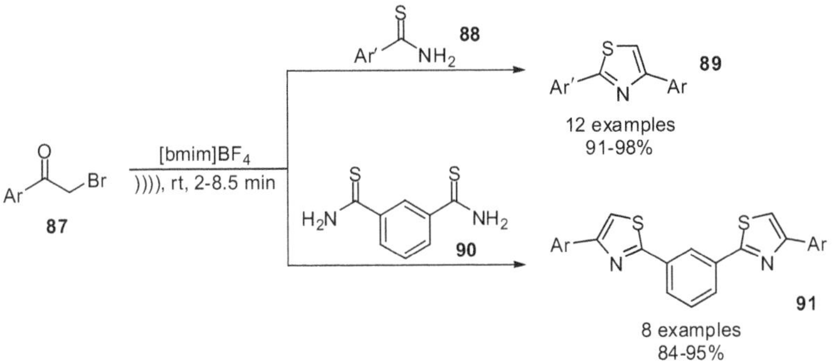
Ar'
NH_2
88
89
12 examples
91-98%
$[bmim]BF_4$
)))), rt, 2-8.5 min
Br
Ar
87
H_2N
NH_2
90
91
8 examples
84-95%

Scheme 25.

Among the natural products containing a 1,3-thiazole ring, thiamine (aneurine, vitamin B_1) is of great importance (Eicher & Hauptmann, 2003). Several 2-(*N*-arylamino)-4-arylthiazoles (**94**) were prepared by the reaction of *a*-bromoacetophenones (**92**) with *N*-aryl substituted thioureas (**93**), as in the classical Hantzsch synthesis, but using ultrasonic irradiation (Scheme 26) (Gupta et al., 2010). This further confirmed that thiazole heterocycles can be conveniently synthesized in good yields (88-97%) by the application of sonochemistry. The insecticidal activity of these 1,3-thiazoles was evaluated.

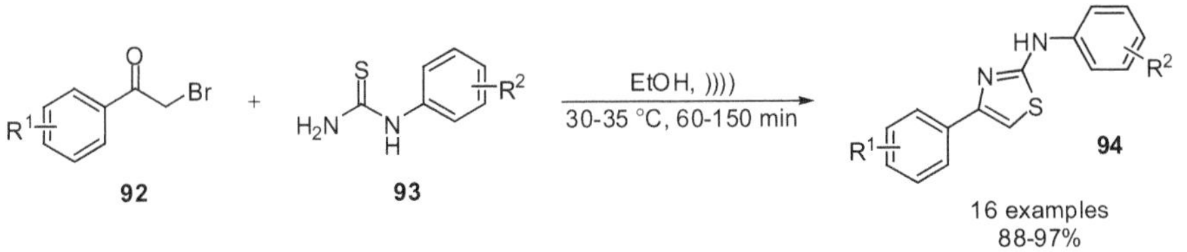
Br
+
H_2N
EtOH,))))
30-35 °C, 60-150 min
92
93
94
16 examples
88-97%

Scheme 26.

Recently, we reported an ultrasound-based procedure for the synthesis of pyrazolyl-substituted thiazoles (**97**) by the cyclization reaction between thiocarbamoyl-pyrazoles (**95**) and *a*-bromoacetophenone (**96**) (Venzke et al., 2011). The reactions occurred in only 15 minutes in ethanol at room temperature, affording the pure products in 47-93% yields by simple filtration of the reaction mixture (Scheme 27).

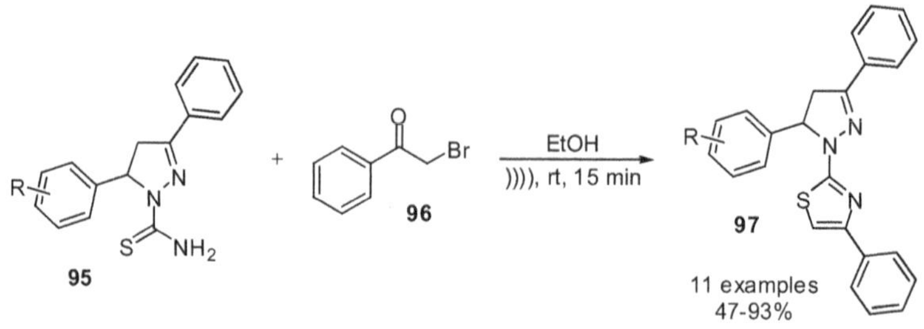

Scheme 27.

Recently, Mamaghani and co-workers described a sonochemical method for the preparation of iminothiazolidinones (**102** and **103**) (Mamaghani et al., 2011). Thioureas (**100**) were generated *in situ* and treated with a mixture of a suitable aldehyde (**101**), chloroform and 1,8-diazabicyclo[5.4.0]undec-7-ene (DBU) in dimethyl ether (DME) under an inert atmosphere. Subsequent addition of aqueous NaOH at 0 °C and sonication furnished the products (**102** and **103**) in 75-91% yields (Scheme 28). A 1:1 mixture of regioisomers was observed when *N*-cyclohexyl-*N'*-ethylthiourea was employed. However, a regiosselective reaction took place with other substituents in the thiourea. The target molecules were obtained in better yields and much shorter reaction times using ultrasound than with conventional methodology.

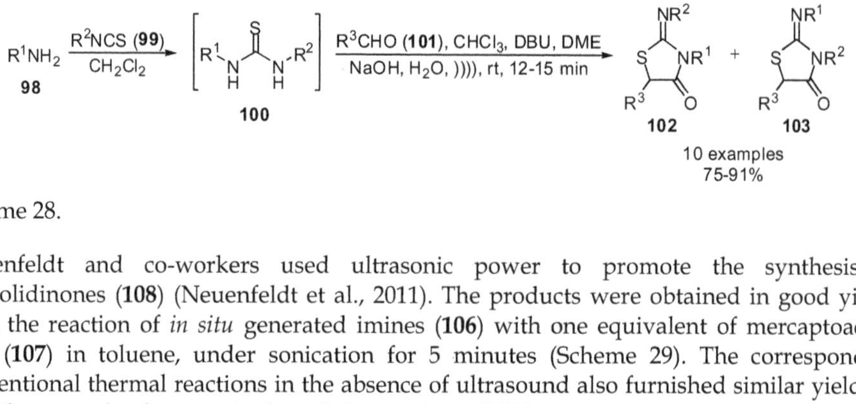

Scheme 28.

Neuenfeldt and co-workers used ultrasonic power to promote the synthesis of thiazolidinones (**108**) (Neuenfeldt et al., 2011). The products were obtained in good yields from the reaction of *in situ* generated imines (**106**) with one equivalent of mercaptoacetic acid (**107**) in toluene, under sonication for 5 minutes (Scheme 29). The corresponding conventional thermal reactions in the absence of ultrasound also furnished similar yields of these heterocycles, but required much longer times (16 h).

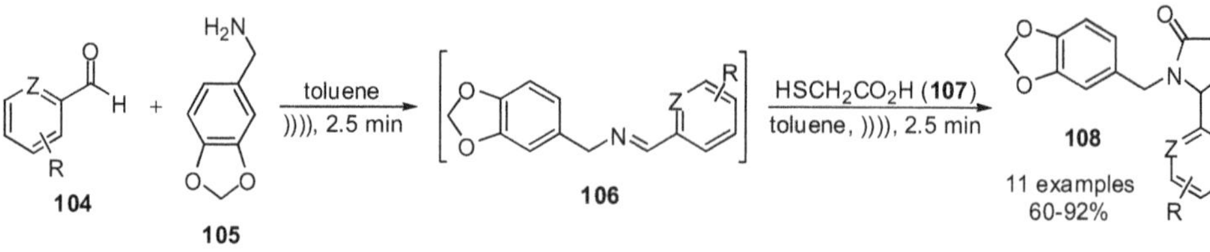

Scheme 29.

A library of spiro[indole-thiazolidinones] (112) was prepared sonochemically by a three component reaction in aqueous medium in the presence of cetyltrimethylammonium bromide (CTAB) as a phase transfer catalyst (Dandia et al., 2011). The reaction of indole-2,3-diones (109), aryl- or heteroaryl-amines (110), and α-mercaptocarboxylic acids (111) under ultrasound for 40-50 minutes afforded the target molecules in good to excellent yields (80-98%) (Scheme 30).

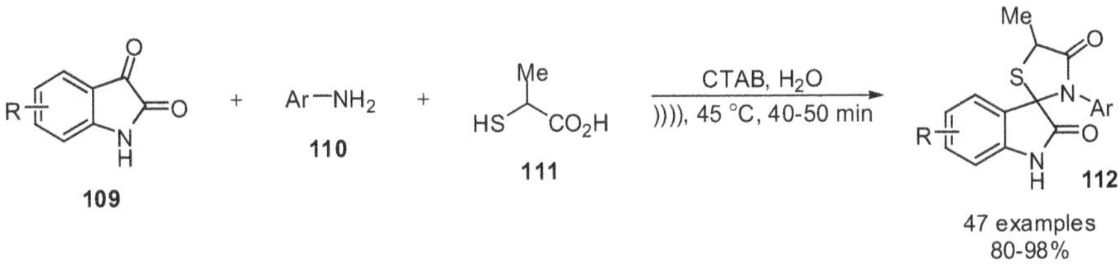

Scheme 30.

3.6 Selenazole derivatives

An ultrasound-mediated preparation of 1,3-selenazoles (115) was reported by Lalithamba and co-workers in 2010 (Lalithamba et al., 2010). The products were efficiently prepared by treatment of bromomethyl ketones (113) with selenourea (114) in acetone under ultrasonic irradiation for 5-10 minutes (Scheme 31).

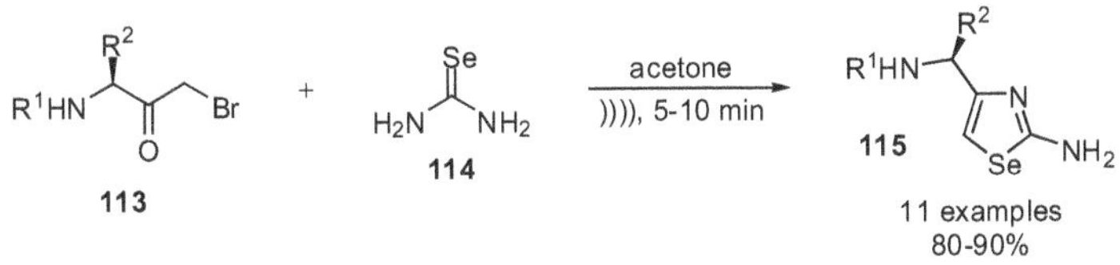

Scheme 31.

4. Azoles with three heteroatoms

4.1 Triazole derivatives

Ultrasonic activation of metal catalysts due to mechanical depassivation has been extensively exploited in organic synthesis. In this context, Cravotto and co-workers (Cravotto et al., 2010) reported an efficient copper-catalyzed azide-alkyne cycloaddition reaction for producing 1,2,3-triazole derivatives (118) using ultrasound or ultrasound and microwave simultaneously. Other activation methods were tested, but the best results were obtained when azides (116) and terminal alkynes (117) were sonicated in the presence of Cu turnings in dioxane/H$_2$O at 70 °C or DMF at 100 °C (Scheme 32). Substitution of water by DMF was required to prevent the formation of copper complexes that made the purification of the products difficult when 6-monoazido-β-cyclodextrin derivatives were utilized as starting materials. No particularly significant differences were observed between the efficiencies of the reactions performed using ultrasound or ultrasound/microwave irradiation.

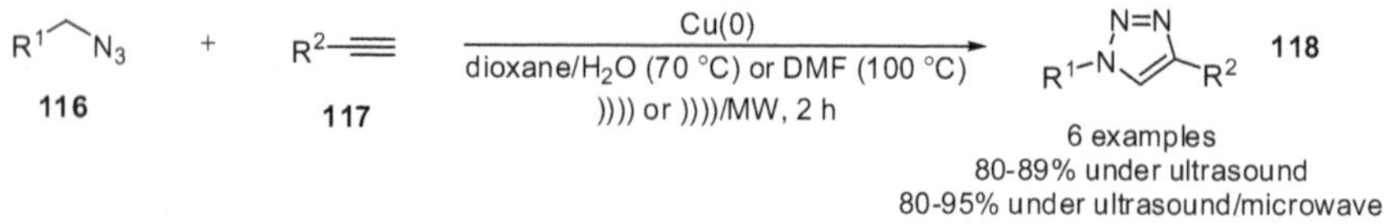

Scheme 32.

4.2 Oxadiazole derivatives

Azoles containing polyhaloalkyl groups are of considerable interest due to their potential herbicidal, fungicidal, insecticidal, analgesic, antipyretic, and anti-inflammatory properties. In addition, 1,2,4-oxadiazoles are reported to posses various types of biological activities (Elguero et al., 2002). Very recently, the rapid preparation of 1,2,4-oxadiazoles (**121**) under ultrasound irradiation was reported (Bretanha et al., 2011). The products were obtained with short reaction times (15 minutes) and in excellent yields (84-98%) (Scheme 33).

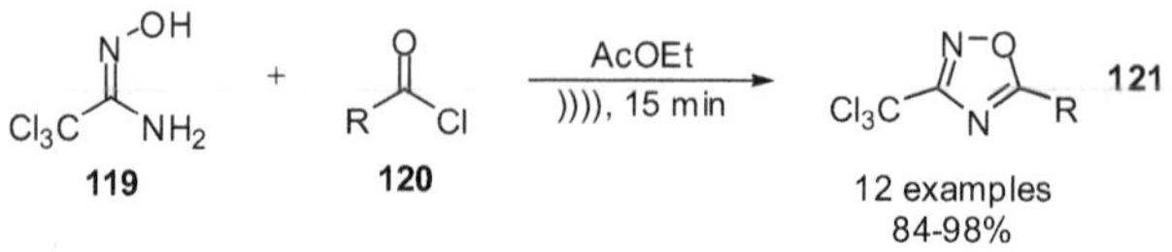

Scheme 33.

4.3 Thiadiazole derivatives

1,3,4-Thiadiazole derivatives (**124**) were synthesized by the reaction of 1-methyl-5-oxo-3-phenyl-2-pyrazolin-4-thiocarboxanilide (**122**) with a series of hydrazonyl halides or N,N'-diphenyl-oxalodihydrazonoyl dichloride (**123**) in the presence of triethylamine (TEA) under ultrasonic irradiation (Scheme 34) (El-Rahman et al., 2009). The products were obtained in excellent yields in short reaction times.

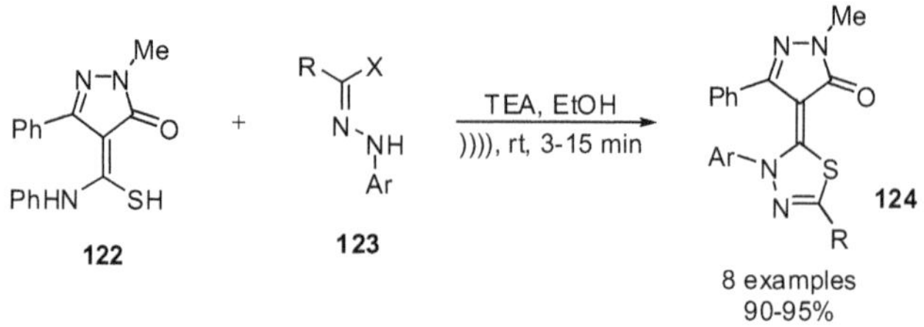

Scheme 34.

5. Azoles with four heteroatoms

5.1 Tetrazole derivatives

In 2010, Chermahini and co-workers reported the clay-catalyzed preparation of tetrazoles (**127**) under ultrasound (Scheme 35) (Chermahini et al., 2010). Compared to conventional heating, ultrasonic irradiation reduced the reaction times and increased the catalyst activity. Unfortunately, the yields obtained by this methodology were not specified by the authors.

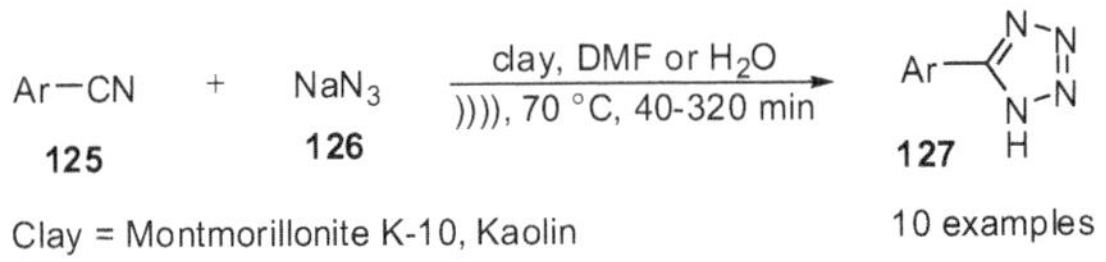

Scheme 35.

6. Miscellaneous

4-Sulphonyl-substituted pyrazoles (130) and isoxazoles (132) were synthesized *via* the one-pot reaction of the carbanions of 1-aryl-2-(phenylsulphonyl)-ethanone (128) with several different hydrazonyl halides (129) or 1-aryl-2-bromo-2-hydroximinoethanones (131) in ethanol, respectively (Scheme 36) (Saleh et al., 2009). These reactions were accelerated by ultrasonic irradiation and the products were isolated in high yields (90-97%).

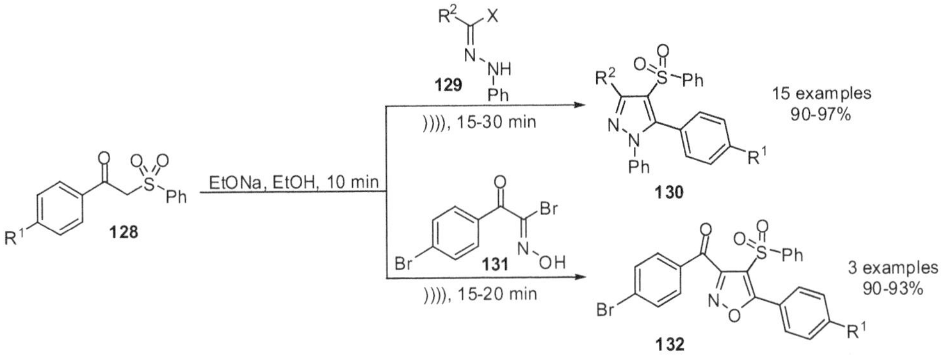

Scheme 36.

In 2009, Al-Zaydi (Al-Zaydi, 2009) reported the synthesis of triazole (135) and pyrazole (138) derivatives starting from arylhydrazononitriles (133) under ultrasonic irradiation (Scheme 37). The triazoles (135) were obtained *via* amidoxime intermediates (134) followed by cyclization with elimination of water. The pyrazoles (138) were prepared directly by reaction with chloroacetonitrile (136). This latter reaction involves the formation of a non-isolable intermediate (137) that undergoes intramolecular cyclization to give the final products (138).

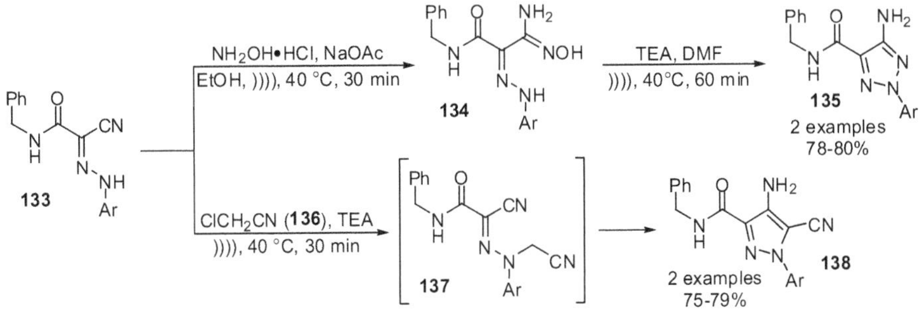

Scheme 37.

In 2009, we described a scaled-up sonochemical method to convert acetylacetone (139) into structurally simple pyrazoles (140) or an isoxazole (141) in aqueous media (Scheme 38) (Silva et al., 2009). The products were obtained after sonication for only 10 minutes, as compared to 12 hours in the thermal reaction without ultrasound.

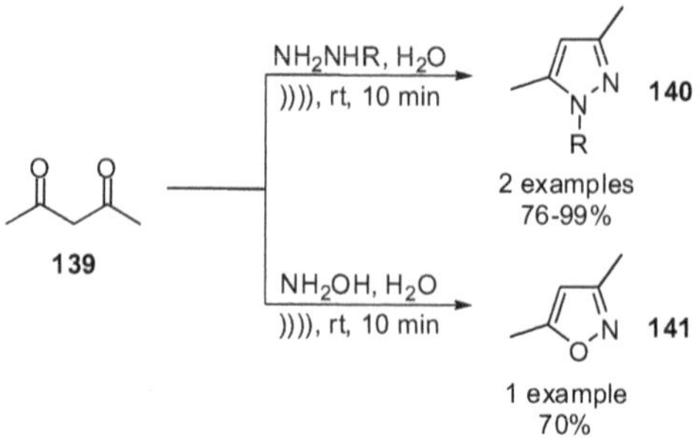

Scheme 38.

As shown in Scheme 39 (Shinde et al., 2010), 2-ethyl-2-methyl-4H-chromen-4-ones (142) were transformed into semicarbazones (143) and thiosemicarbazones (146). The semicarbazones (143) could be sonochemically converted into selenadiazole derivatives (144) in 30 minutes by treatment with SeO_2 in acetic anhydride. The same semicarbazones (143) afforded 1,2,3-thiadiazoles (145) under ultrasonic irradiation in 20 minutes in the presence of thionyl chloride. Similarly, sonication of the thiosemicarbazones (146) for 45 minutes in acetic anhydride produced the thiadiazolines (147) in good yields. Comparison of these results with those for the same reactions under microwave irradiation showed that the time required was longer and yields lower.

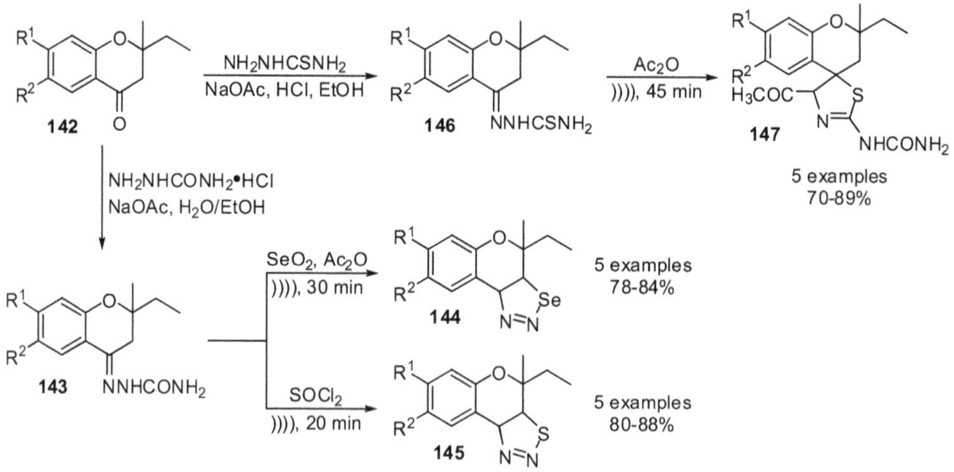

Scheme 39.

The syntheses of 1,3,4-thiadiazoles (149) and 1,3,4-triazoles (150) (Scheme 40) via the intramolecular cyclocondensation of benzofuran-substituted thiosemicarbazides (148) in acidic or basic media, respectively, has been reported (Shinde et al., 2010). The reactions were carried out employing ultrasound, microwaves and conventional conditions. Ultrasound afforded the best yields. The scope of these reactions was subsequently expanded by Shelke and co-workers (Scheme 41) (Shelke et al., 2010).

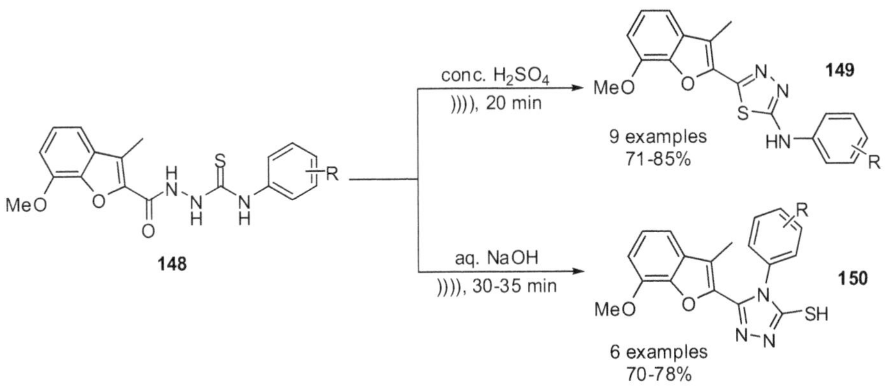

Scheme 40.

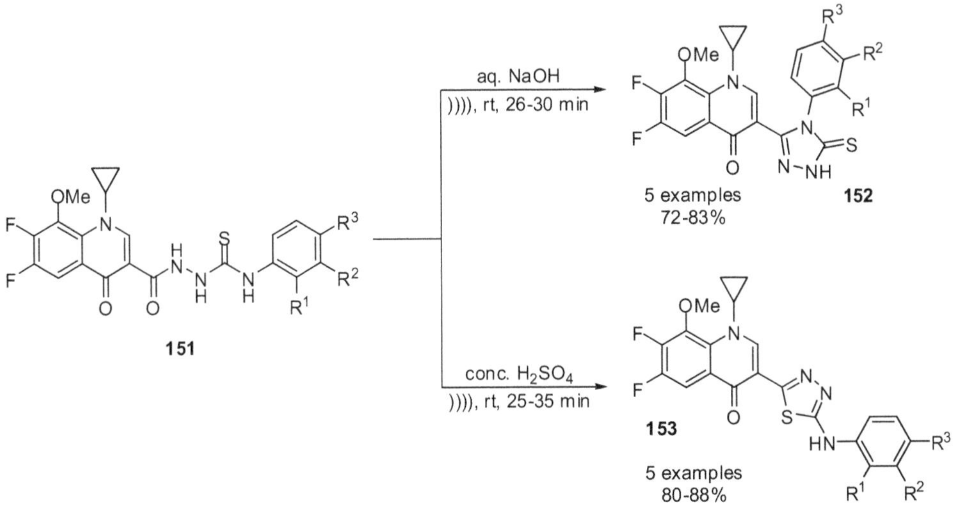

Scheme 41.

Yuan and Guo (Yuan & Guo, 2011) reported the one-pot cyclocondensation of *o*-aminothiophenol (155) or aromatic *o*-diamines (157) with aromatic aldehydes (154) in the presence of chlorotrimethylsilane [TMSCl/Fe(NO$_3$)$_3$] in dimethylformamide under ultrasonic irradiation for the preparation of benzothiazoles (156) and benzimidazoles (158) in 84-97% yields (Scheme 42).

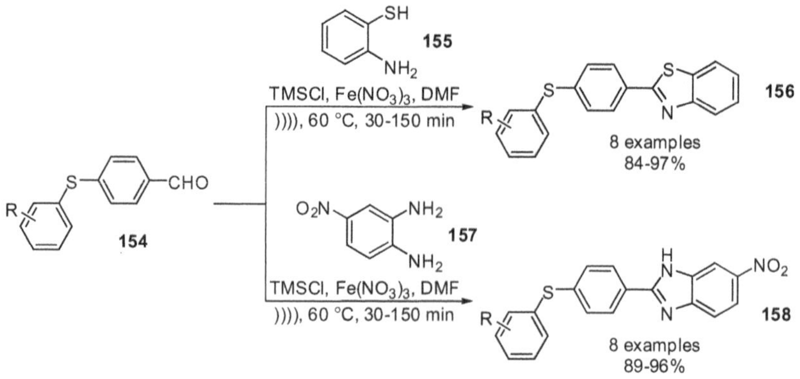

Scheme 42.

7. Conclusion

As we have shown in this chapter, several convenient ultrasound-promoted synthetic methodologies have been established for the preparation of the title class of compounds. The main advantages of the use of ultrasound in azole synthesis are evident when compared with classical methodologies i.e., a reduction in the reaction times and an improvement in yields. Most of the papers covered by this review employed simple ultrasonic cleaning baths as energy sources. Although these low potency sources of ultrasonic radiation are usually less efficient than immersion sonication probes, requiring longer reaction times, cleaning baths are relatively inexpensive and widely available in chemistry laboratories.

8. Acknowledgement

The authors acknowledge CAPES and the CNPq (INCT Estudos do Meio Ambiente, grant 573.667/2008-0) for the financial support. FHQ is affiliated with INCT-Catalysis and NAP-PhotoTech (the USP Research Consortium for Photochemical Technology) and thanks the CNPq for fellowship support.

9. References

Ahmad, M.; Siddiqui, H. L.; Zia-ur-Rehman, M. & Parvez, M. (2010). Anti-oxidant and anti-bacterial activities of novel N'-arylmethylidene-2-(3,4-dimethyl-5,5-dioxidopyrazolo[4,3-c][1,2]benzothiazin-2(4H)-yl) acetohydrazides. *European Journal of Medicinal Chemistry*, Vol. 45, pp. 698-704.

Al-Mutairi, A. A.; El-Baih, F. E. M. & Al-Hazimi, H. M. (2010). Microwave versus ultrasound assisted synthesis of some new heterocycles based on pyrazolone moiety. *Journal of Saudi Chemical Society*, Vol. 14, pp. 287-299.

Al-Zaydi, K. M. (2009). A simplified green chemistry approaches to synthesis of 2-substituted 1,2,3-triazoles and 4-amino-5-cyanopyrazole derivatives conventional heating versus microwave and ultrasound as ecofriendly energy sources. *Ultrasonics Sonochemistry*, Vol. 16, pp. 805-809.

Arani, N. M. & Safari, V. J. (2011). A rapid and efficient ultrasound-assisted synthesis of 5,5-diphenylhydantoins and 5,5-diphenyl-2-thiohydantoins. *Ultrasonics Sonochemistry*, Vol. 18, pp. 640-643.

Bretanha, L. C.; Teixeira, V. E.; Ritter, M.; Siqueira, G. M.; Cunico, W.; Pereira, C. M. P. & Freitag, R. A. (2011). Ultrasound-promoted synthesis of 3-trichloromethyl-5-alkyl(aryl)-1,2,4-oxadiazoles. *Ultrasonics Sonochemistry*, Vol. 18, pp. 704-707.

Cella, R. & Stefani, H. A. (2009). Ultrasound in heterocycles chemistry. *Tetrahedron*, Vol. 65, pp. 2619-2641.

Chermahini, A. N.; Teimouri, A.; Momenbeik, F.; Zarei, A.; Dalirnasab, Z.; Ghaedi, A. & Roosta, M. (2010). Clay-catalyzed synthesis of 5-substituent 1*H*-tetrazoles. *Journal of Heterocyclic Chemistry*, Vol. 47, pp. 913-922.

Cintas, P. & Luche, J.-L. (1999). Green chemistry: the sonochemical approach. *Green Chemistry*, pp. 115-125.

Cravotto, G. & Cintas, P. (2006). Power ultrasound in organic synthesis: moving cavitational chemistry from academia to innovative and large-scale applications. *Chemical Society Reviews*, Vol. 35, pp. 180-196.

Cravotto, G.; Fokin, V. V.; Garella, D.; Binello, A.; Boffa, L. & Barge, A. (2010). Ultrasound-promoted copper-catalyzed azide-alkyne cycloaddition. *Journal of the Combinatorial Chemistry*, Vol. 12, pp. 13-15.

Dandia, A.; Singh, R.; Bhaskaran, S. & Samant, S. D. (2011). Versatile three component procedure for combinatorial synthesis of biologically relevant scaffold spiro[indole-thiazolidinones] under aqueous conditions. *Green Chemistry*, Vol. 13, pp. 1852-1859.

Eicher, T. & Hauptmann, S. (2003), In *The Chemistry of Heterocycles, Second Edition.* Wiley-VCH, Saarbrücken e Leipzig, pp. 154.

Elguero, J.; Goya, P.; Jagerovic, N.; Silva, A.M.S. (2002). In *Targets in Heterocyclic System,* Italian Society of Chemistry, Vol. 6, pp. 167.

El-Rahman, N. M. A.; Saleh, T. S. & Mady, M. F. (2009). Ultrasound assisted synthesis of some new 1,3,4-thiadiazole and bi(1,3,4-thiadiazole) derivatives incorporating pyrazolone moiety. *Ultrasonics Sonochemistry*, Vol. 16, pp. 70-74.

Fry, A. J. & Herr, D. (1978). Reduction of α,α'-dibromo ketones by ultrasonically dispersed mercury in protic solvents. *Tetrahedron Letters*, Vol. 19, pp. 1721-1724.

Gupta, R.; Gupta, N. & Jain, A. (2010). Improved synthesis of chalcones and pyrazolines under ultrasonic irradiation. *Indian Journal of Chemistry*, Vol. 49B, pp. 351-355.

Gupta, R.; Sharma, D. & Singh, S. (2010). Eco-friendly synthesis and insecticidal activity of some fluorinated 2-(N-arylamino)-4-arylthiazoles. *Phosphorus, Sulfur, and Silicon*, Vol. 185, pp. 1321-1331.

Joshi, R. S.; Mandhane, P. G.; Dabhade, S. K. & Gill, C. H. (2010). Tetrabutylammonium fluoride (TBAF) catalysed synthesis of 2-arylbenzimidazole in water under ultrasound irradiation. *Journal of the Chinese Chemical Society*, Vol. 57, pp. 1227-1231.

Kanagarajan, V.; Ezhilarasi, M. R. & Gopalakrishnan, M. (2011). 'One-pot' ultrasound irradiation promoted synthesis and spectral characterization of an array of novel 1,1'-(5,5'-(1,4-phenylene) *bis*(3-aryl-1*H*-pyrazole-5,1(4*H*,5*H*)-diyl))diethanones, a *bis* acetylated pyrazoles derivatives. *Spectrochimica Acta Part A*, Vol. 78, pp. 635-639.

Lalithamba, H. S.; Narendra, N.; Naik, S. A. & Sureshbabu, V. V. (2010). Ultrasound mediated synthesis of 2-amino-1,3-selenazoles derived from Fmoc/Boc/Z-a-amino acids. *Arkivoc*, Vol. ix, 77-90.

Li, J.-T.; Liu, X.-R. & Sun, M.-X. (2010). Synthesis of glycoluril catalyzed by potassium hydroxide under ultrasound irradiation. *Ultrasonics Sonochemistry*, Vol. 17, pp. 55-57.

Li, J.-T.; Yin, Y.; Li, L. & Sun, M.-X. (2010). A convenient and efficient protocol for the synthesis of 5-aryl-1,3-diphenylpyrazole catalyzed by hydrochloric acid under ultrasound irradiation. *Ultrasonics Sonochemistry*, Vol. 17, pp. 11-13.

Li, J.-T.; Zhai, X.-L. & Chen, G.-F. (2010). Ultrasound promoted one-pot synthesis of 3-aza-6,10-diaryl-2-oxa-spiro[4.5]decane-1,4,8-trione. *Ultrasonics Sonochemistry*, Vol. 17, pp. 356-358.

Luche, J.-L. & Damiano, J. C. (1980). Ultrasound in organic synthesis. 1. Effect on the formation of lithium organometallic reagents. *Journal of the American Chemical Society*, Vol. 102, pp. 7926-7927.

Machado, P.; Lima, G. R.; Rotta, M.; Bonacorso, H. G.; Zanatta, N. & Martins, M. A. P. (2011). Efficient and highly regioselective synthesis of ethyl 1-(2,4-dichlorophenyl)-1*H*-pyrazole-3-carboxylates under ultrasound irradiation. *Ultrasonics Sonochemistry*, Vol. 18, pp. 293-299.

Mamaghani, M. & Dastmard, S. (2009). One-pot easy conversion of Baylis-Hillman adducts into arylpyrazoles under ultrasound irradiation. *Arkivoc*, Vol. ii, pp. 168-173.

Mamaghani, M.; Loghmanifar, A. & Taati, M. R. (2011). An efficient one-pot synthesis of new 2-imino-1,3-thiazolidin-4-ones under ultrasonic conditions. *Ultrasonics Sonochemistry*, Vol. 18, pp. 45-48.

Mason, T. J. (1997). Ultrasound in synthetic organic chemistry. *Chemical Society Reviews*, Vol. 26, pp. 443-451.

Mason, T. J. & Lorimer, J. P. (2002). *Applied sonochemistry: uses of power ultrasound in chemistry and processing*, Wiley-VCH, ISBN 3-527-3020500, Weinheim.

Mirkhani, V.; Moghadam, M.; Tangestaninejad, S.; Mohammadpoor-Baltork, I. & Mahdavi, M. (2009). Preparation of an improved sulfonated carbon-based solid acid as a novel, efficient, and reusable catalyst for chemo selective synthesis of 2-oxazolines and bis-oxazolines. *Monatshefte für Chemie*, Vol. 140, pp. 1489-1494.

Moghadam, M.; Mirkhani, V.; Tangestaninejad, S.; Mohammadpoor-Baltork, I. & Kargar, H. (2009). InCl₃ as an efficient catalyst for synthesis of oxazolines under thermal, ultrasonic and microwave irradiations. *Journal of the Iranian Chemical Society*, Vol. 6, pp. 251-258.

Nabid, M. R.; Rezaei, S. J. T.; Ghahremanzadeh, R. & Bazgir, A. (2010). Ultrasound-assisted one-pot, three-component synthesis of 1*H*-pyrazolo[1,2-*b*]phthalazine-5,10-diones. *Ultrasonics Sonochemistry*, Vol. 17, pp. 159-161.

Neuenfeldt, P. D.; Duval, A. R.; Drawanz, B. B.; Rosales, P. F.; Gomes, C. R. B.; Pereira, C. M. P. & Cunico, W. (2011). Efficient sonochemical synthesis of thiazolidinones from piperonilamine. *Ultrasonics Sonochemistry*, Vol. 18, pp. 65-67.

Noei, J. & Khosropour, A. R. (2009). Ultrasound-promoted a green protocol for the synthesis of 2,4-diarylthiazoles under ambient temperature in [bmim]BF₄. *Ultrasonics Sonochemistry*, Vol. 16, pp. 711-717.

Pathak, V. N.; Joshi, R.; Sharma, J.; Gupta, N. & Rao, V. M. (2009). Mild and ecofriendly tandem synthesis, and spectral and antimicrobial studies of N^1-acetyl-5-aryl-3-(substituted styryl)pyrazolines. *Phosphorus, Sulfur, and Silicon*, Vol. 184, pp. 1854-1865.

Pizzuti, L.; Martins, P. L. G.; Ribeiro, B. A.; Quina, F. H.; Pinto, E.; Flores, A. F. C.; Venzke, D. & Pereira, C. M. P. (2010). Efficient sonochemical synthesis of novel 3,5-diaryl-4,5-dihydro-1H-pyrazole-1-carboximidamides. *Ultrasonics Sonochemistry*, Vol. 17, pp. 34-37.

Pizzuti, L.; Piovesan, L. A.; Flores, A. F. C.; Quina, F. H. & Pereira, C. M. P. (2009). Environmentally friendly sonocatalysis promoted preparation of 1-thiocarbamoyl-3,5-diaryl-4,5-dihydro-1H-pyrazoles. *Ultrasonics Sonochemistry*, Vol. 16, pp. 728-731.

Richards, W. T. & Loomis, A. L. (1927). The chemical effects of high-frequency sound waves. I. A preliminary survey. *Journal of the American Chemical Society*, Vol. 49, pp. 3086-3100.

Rodrigues-Santos, C. E. & Echevarria, A. (2011). Convenient syntheses of pyrazolo[3,4-b]pyridin-6-ones using either microwave or ultrasound irradiation. *Tetrahedron Letters*, Vol. 52, pp. 336-340.

Saleh, T. S. & El-Rahman, T. S. (2009). Ultrasound promoted synthesis of substituted pyrazoles and isoxazoles containing sulphone moiety. *Ultrasonics Sonochemistry*, Vol. 16, pp. 237-242.

Sant'Anna, G. S.; Machado, P.; Sauzem, P. D.; Rosa, F. A.; Rubin, M. A.; Ferreira, J.; Bonacorso, H. G.; Zanatta, N. & Martins, M. A. P. (2009). Ultrasound promoted synthesis of 2-imidazolines in water: a greener approach toward monoamine oxidase inhibitors. *Bioorganic & Medicinal Chemistry Letters*, Vol. 19, pp. 546-549.

Shekouhy, M. & Hasaninejad, A. (2012). Ultrasound-promoted catalyst-free one pot four component synthesis of 2H-indazolo[2,1-b]phthalazine-triones in neutral ionic liquid 1-butyl-3-methylimidazolium bromide. *Ultrasonics Sonochemistry*, Vol. 19, pp. 307-313.

Shelke, K. F.; Sapkal, S. B.; Sonar, S. S.; Madje, B. R.; Shingate, B. B. & Shingare, M. S. (2009). An efficient synthesis of 2,4,5-triaryl-1H-imidazole derivatives catalyzed by boric acid in aqueous media under ultrasound-irradiation. *Bulletin of the Korean Chemical Society*, Vol. 30, pp. 1057-1060.

Shelke, S.; Mhaske, G.; Gadakh, S. & Gill, C. (2010). Green synthesis and biological evaluation of some novel azoles as antimicrobial agents. *Bioorganic & Medicinal Chemistry Letters*, Vol. 20, pp. 7200-7204.

Shinde, A. D.; Kale, B. Y.; Shingate, B. B. & Shingare, M. S. (2010). Synthesis and characterization of 1-benzofuran-2-yl thiadiazoles, triazoles and oxadiazoles by conventional and non-conventional methods. *Journal of the Korean Chemical Society*, Vol. 54, pp. 582-588.

Shinde, A. D.; Sonar, S. S.; Shingate, B. B. & Shingare, M. S. (2010). Synthesis and biological screening of novel thiadiazoles, selenadiazoles, and spirocyclic benzopyran by ultrasonic and microwave irradiation. *Phosphorus, Sulfur, and Silicon*, Vol. 185, pp. 1594-1603.

Silva, F. A. N.; Galluzzi, M. P.; Albuquerque, B.; Pizzuti, L.; Gressler, V.; Rivelli, D. P.; Barros, S. B. M. & Pereira, C. M. P. (2009). Ultrasound irradiation promoted large-scale preparation in aqueous media and antioxidant activity of azoles. *Letters in Drug Design & Discovery*, Vol. 6, pp. 323-326.

Tiwari, V.; Parvez, A. & Meshram, J. (2011). Benign methodology and improved synthesis of 5-(2-chloroquinolin-3-yl)-3-phenyl-4,5-dihydroisoxazoline using acetic acid aqueous solution under ultrasound irradiation. *Ultrasonics Sonochemistry*, Vol. 18, pp. 911-916.

Venzke, D.; Flores, A. F. C.; Quina, F. H.; Pizzuti, L. & Pereira, C. M. P. (2011). Ultrasound promoted greener synthesis of 2-(3,5-diaryl-4,5-dihydro-1H-pyrazol-1-yl)-4-phenylthiazoles. *Ultrasonics Sonochemistry*, Vol. 18, pp. 370-374.

Yuan, Y.-Q. & Guo, S.-R. (2011). TMSCl/Fe(NO$_3$)$_3$-Catalyzed synthesis of 2-arylbenzothiazoles and 2-arylbenzimidazoles under ultrasonic irradiation. *Synthetic Communications*, Vol. 41, pp. 2169-2177.

Zang, H.; Su, Q.; Mo, Y.; Cheng, B.-W. & Jun, S. (2010). Ionic liquid [emim]OAc under ultrasonic irradiation towards the first synthesis of trisubstituted imidazoles. *Ultrasonics Sonochemistry*, Vol. 17, pp. 749-751.

Zou, Y.; Wu, H.; Hu, Y.; Liu, H.; Zhao, X.; Ji, H. & Shi, D. (2011). A novel and environment-friendly method for preparing dihydropyrano[2,3-c]pyrazoles in water under ultrasound irradiation. *Ultrasonics Sonochemistry*, Vol. 18, pp. 708-712.

6

New Green Oil-Field Agents

Arkadiy Zhukov and Salavat Zaripov
R&D Center, GC «Mirrico»
Russian Federation

1. Introduction

The current state of environmental conditions on planet Earth is a substantial basis for modification as the cleaning and disposal of waste and emissions, and fundamental changes in processes and technology industries. Recently, the transition from administrative methods required to control unwanted emissions and destroy formed by chemical processes harmful substances to a fundamentally different method - the method of green chemistry. Green chemistry in its best incarnation - is an art form that allows not just to get the right stuff, but ideally to get it in a way, which does not harm the environment at any stage of production. Of course, the substance itself must also be friendly to biosphere.

Like any perfected motion requires less force for its implementation and use of methods of green chemistry leads to a reduction of production costs, if only because they do not need to enter the stage of destruction and recycling of hazardous by-products, used solvents and other wastes, as they are simply not formed. Reducing the number of stages leading to energy savings, and this is also a positive impact on environmental and economic assessment of the production. It is important to note that the view of ongoing research from the viewpoint of green chemistry can be useful in purely scientific terms. Often, such a change of paradigm allows scientists to see their own research in a new light and open up new opportunities that benefit science in general.

2. How to implement "green" process

Introduction is a difficult task even for developed countries. In Britain, for example, is now a strongly encouraged interaction between scientists and chemical technologists. Even a joint center for the introduction of «green chemistry» has been created. At the University of Nottingham for the first time in the world to teach a course on green chemistry for chemistry students and chemists and technologists of the last school year. Undergraduates are taught to consider the chemical process as a whole and not fragmented. It is not enough that the specialist can choose the traditional or the most expensive reagent for the industrial synthesis, it is necessary to keep in mind the entire process from beginning to end. The primary sources of the initial reagent (extracted or renewable) as the reagent prepared, nuclear reaction efficiency, solvents, minimizing their use or non-toxic solvents, the selectivity of output allows the costs of by-products to be at a low level, what will ensure the viability of the process.

3. Oil production

Modern man can not exist without the consumption of large amounts of energy. Historical progress of the international community was determined by first of all that mankind has managed to use for practical purposes fossil fuels: coal, oil and natural gas. In the 60s of last century, about three-quarters of world consumption of fuel wood and covered with vegetable substitutes, almost a quarter - coal. The share of oil and gas accounted for about 1%. At the end of the century came the "era of coal." In 1900, its share in energy balance (FEB), the world has risen to 57%, the share of oil and gas was at 2.3% and 0.9%.

Until the mid-nineteenth century, oil was extracted in small amounts (2-5 tons per year) from shallow wells near its natural outlets to the surface. The Industrial Revolution, based on extensive use of steam engines, determined the broad demand for lubricants and light sources (kerosene). There was an increased demand for oil. With the introduction in the late 60-ies, oil well drilling global oil production tenfold increased from 2 to 20 million tonnes by the end of the nineteenth century. In 1900, oil was extracted in 10 countries: Russia, U.S., Dutch East India, Romania, Austria-Hungary, India, Japan, Canada, Germany, Peru. Almost half of total world oil production goes to Russia (9927 tonnes) and USA (8334 tonnes). Throughout the twentieth century. global oil consumption continued to grow rapidly.

The effectiveness of development of oil reduces the loss of its reserves in the subsoil, as not all stocks can be learned for technological reasons: falling well production rates (at least the fall of the reservoir pressure), water content and total depletion of deposits, etc. In order to increase production using different methods of influencing the oil reservoirs. So, in 1994, with the help of thermal, chemical, gas and other methods to increase the recovery factor in the world received an additional 93.4 million tons of oil. That same year, the application of new methods of enhanced oil recovery in the U.S. has allowed to obtain an additional 34.89 million tons, Russia - 11.55 million tonnes of oil. On Romashkino field through the use of enhanced oil recovery techniques more oil in 1993 was 26% with respect to all oil from the field in the same year. However, in recent years the average rate of oil recovery in Russia as a whole on all deposits decreased. Now in more than four fifths of all recoverable oil wells are produced fluids with electric-submersible pumps and gas lift.

4. Environmental issues in oil and gas sector

Oil and gas industry has a negative impact on all components of the environment, especially in northern areas. During the drilling of wells, construction of buildings and land communications in areas of permafrost after the integrity of the protective vegetative cover increases the heating of the soil to great depth. Thermokarst processes cause melting of underground ice. Because of this, the formation of the earth surface subsidence, deep channels, gullies, formation of new lakes, marshes and valleys, which in turn increases the likelihood of deformation of the pipe and breaks. In the process of production, preparation, transportation, storage, processing and utilization of oil and gas produces toxic chemicals. Contain their bowels pollutes the waste water, natural landscapes and water bodies. Thus, in the West Siberian petroleum province in many areas of Ob river and its tributaries, the content of organic pollutants exceeds the MPC by ten times. The volume of water consumed by a drilling rig, for example, in the gas industry, ranging from 25 to 120 m3 per day. Daily volume of wastewater generated is 20-40 m3 per well. The annual volume of drilling wastewater averaging 777 thousand m3, including cuttings 221.9 thousand m3. Significant damage to aquatic ecosystems causing a flowing drilling and drilling fluid from the barns, accidents at oil pipelines and fishing facilities. Accidents on main oil and gas pipelines have a catastrophic impact on the environment. However, most of the Russian pipelines operated by longer regulatory period, it makes them targets of environmental risk. Thus, in Russia on gas pipelines from 1960 to 1990, there were 1,200 accidents. The accident at the Western Siberian oil product in 1989 led to a major train accident, accompanied by a powerful explosion, fire, and many large-scale casualties. In 1994, as a result of an accident on an oil pipeline in the Komi Republic were contaminated with large surface area and significant volumes of oil got into streams and rivers. Oil and gas facilities emit greenhouse gases, nitrogen oxides, sulfur dioxide and other toxic and natural hydrocarbons themselves.

The structure of the oil is changed for the worse. Despite the advances in technology, methods, Geophysics, effort and investments aimed at exploration, a large-scale increase in lung stocks are not observed, but on the contrary, the future of oil production is associated with heavy oil onshore, offshore production (including the Arctic) and production of deep-sea oil. Russian oil and gas producers are is the second largest country after Saudi Arabia. Currently in Russia there is a problem structure of proven oil reserves: the current oil reserves, the share of hard-to-oil exceeded 60%. In this regard, increased production of heavy oil: in 2005, was produced 42.5 million tons heavy and extra heavy oil. These are hard-to-oil, or mode of occurrence or the quality of raw materials. This category should also include most of the oil reserves in undergas deposits. If to this be added to the extraction yield in underdeveloped areas with difficult climatic conditions and the near absence of economic and transport infrastructure, the production could be on the verge of economic efficiency. To some extent, this is a world problem. The deterioration of the structure of inventories in the future will inevitably affect the price rise in oil

United States opened for the development of offshore deposits of the Pacific, Atlantic oceans, which have long been "locked up" officially "for environmental reasons." But after a major accident in the Gulf of Mexico in April 2010 on a platform of deep «Deep Horizon», Congress was forced to ban the further development of offshore fields until the development of regulations safety. There is no doubt that in the near future, the ban will be lifted, but the fact that the offshore fields have been opened for development, said that other more lucrative prospects of large-scale production in the U.S. has no land. In addition, in itself a step toward increasing production of deep oil suggests that the U.S. government are convinced that oil prices will remain high in the future, i.e. the proportion of hard-to-oil in world production will increase.

Renewable energy projects are actively promoted. The program of development of wind, solar and biofuels are set up all around the world. The leaders are the United States, China and the European Union. For example in Europe in 2020 is planned to increase the share of renewable energy to 20%. Officially this is because "the reduction of human impact on climate change and reducing emissions of CO2 in the atmosphere." But in order for these programs earned, and renewable energy has become competitive, you need high prices for oil and gas. A rise in prices, in turn, negatively affects economic growth. With all that research on human impacts on climate are controversial, as there was much talk, so to some extent is illogical rush to develop renewable energy (with the creation of objective difficulties for the economy). On the other hand, if we take into account the possibility of reducing oil production in the next two decades, this increased interest in renewable energy is a logical and understandable.

5. Oilfield chemicals

Significant branch of the modern chemical industry is the manufacture of products used in the processes of the oil and oil transportation. Each year the requirements for this kind of reagents increasingly tightened, for example, to decrease the volume of product flow while increasing its efficiency. Below we will consider some classes of oilfield chemicals and provide examples of available commercial products, including meeting the requirements of «green chemistry».

5.1 Demulsifiers

In recent years, many deposits are opened in the late stage of development which has significant water content of the output.

As a result, commercial facilities pose serious technological challenges associated with the need to handle large quantities of water extracted simultaneously. Formation of emulsions during oil production is the main cause for large losses of oil, cost of transportation and preparation for recycling.

With increasing water content in oil at 1% of transportation costs increase by 3-5% for each transfer. In addition to costs directly in the oil industry, large volumes of water extracted along the way during transport cause the destruction of oilfield corrosion and environmental problems due to accidents of pipelines. Usually, the destruction of oil-water emulsion is heated addition of demulsifiers of this mixture.

Demulsifiers are currently anion (cation) active and nonionic surfactants: the block copolymers of ethylene oxide and propylene oxide, ethoxylated amines, higher fatty alcohols, alkylphenols, etc. Even with relatively low consumption of reagents (40-100 g/t) rather acute problem recycling of surface-active substances produced by many thousands of tons of the modern chemical industry.

Creating a "green" brands demulsifiers is justified not only environmental, but also with the economic position as a biodegradable agent does not require, or at least reduces the cost of cleanup and disposal of waste containing it. So do not consider the desire to create an "environmentally friendly" chemicals like fashion - it can actually lead to significant cost savings.

Here are some examples of such demulsifiers.

On the basis of wood chemical product manufacturing - tall oil, which consists of fatty acids (oleic, linoleic, palmitic) and resin acids: abietic, neoabietic, digidroabietic (Figure 1) by their ethoxylation demulsifiers were obtained, but tests showed their lower efficiency compared to demulsifier obtained ethoxylation of fatty acids.

On the basis of xylitol (derived from waste cotton plants, Figure 2) and synthetic fatty acids was obtained nonionic demulsifier, which showed a result of industrial tests of high efficiency.

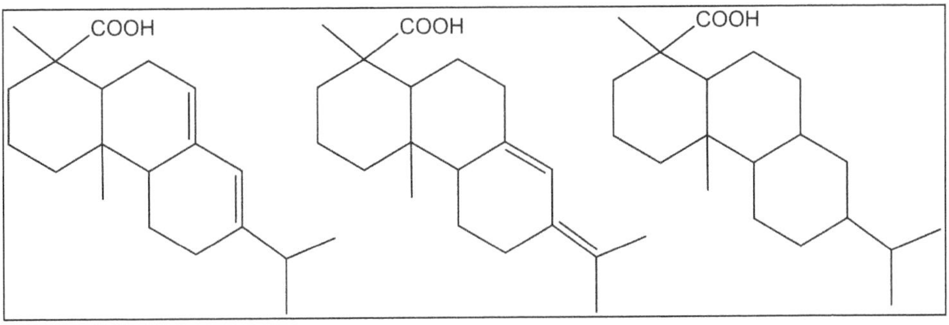

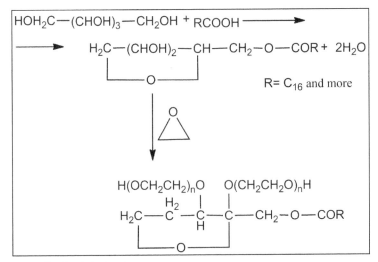

Fig. 1. Abietic acids

Fig. 2. Synthesis of xylitol derivatives

Promising "green" platform for demulsifiers is glycerol. Ethoxylation its esters (acetate, stearate), a number of new demulsifiers.

Despite the high efficiency of demulsifiers on the basis of alkyl phenols, their use should be limited due to toxicity of phenolic fragment.

It was found that emulsions are efficient destroyers of silicone derivatives that do not contain toxic groups, such as phenolic or aryl. Such compounds are promising for use in oil field chemistry.

Leading manufacturers of oilfield chemicals have in their product line, "green" demulsifiers, for example, Clariant offers a similar product under the brand «Phasetreat». Are patented cross-linked esters of glycerol and pentaerythritol.

There are patents in which as demulsifiers proposed use oligoglicosides containing alkyl radicals and oxyethylene-, oxypropylene-substituents. Specially negotiated their biodegradability.

Demulsifier, patented Nalco (Figure 3) is not only biodegradable but also non-toxic compound.

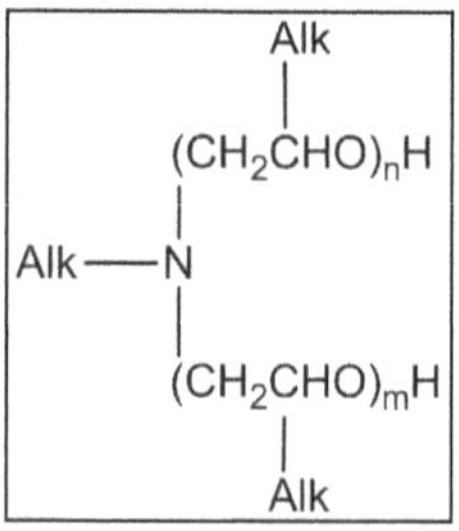

$$Alk-N\begin{cases}(CH_2CHO)_nH \cdot Alk \\ (CH_2CHO)_mH \cdot Alk\end{cases}$$

Fig. 3. Nalco' s demulsifier

The new direction is the use of dendrimers as demulsifiers, polyesters such as oxyalkylated Boltorn H20 and similar compounds (Figure 4).

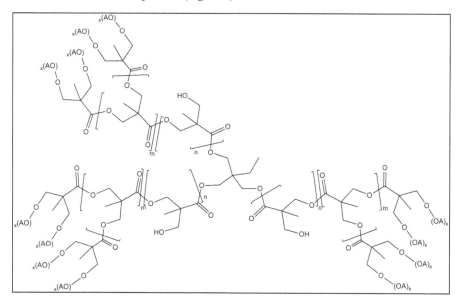

Fig. 4. Dendritic demulsifier

Despite the apparent complexity of its synthesis is rather simple.

Among the macromolecular platforms for the synthesis of interest are the macrocycles, such as calix[4]arenes. The possibility of fixing the macrocyclic rings in several spatial configurations, which provides a different position of substituents in space relative to each other (Figure 5), in combination with non-toxicity and ease of chemical modification of the lower rim of calixarenes makes promising molecular platforms for the creation on their basis of various reagents. Although many authors noted non-toxic derivatives of calixarene, yet the question of their biodegradability and metabolism studied enough; phenolic fragments, forming a skeleton macromolecules suggest that the degradation products could be toxic substances.

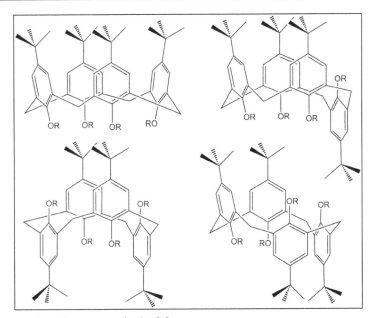

Fig. 5. The spatial configuration of calix [4] arene

The vast majority of demulsifiers are oxyalkylated derivatives (poly-) alcohols, acids and amines. Search and development of new products should be the rational choice of the starting material - the basis platform for oxyalkylation. Careful attention should be paid to renewable raw materials and waste paper and pulp, wood chemistry and food industry.

In addition to chemical methods of destruction of water-oil emulsions to improve and develop new techniques based on physical effects: heat, magnetic field, etc., as well as use the combined, integrated approach - the use of demulsifiers in conjunction with the hydrodynamic effects and magnetic devices.

5.2 Corrosion inhibitors

To date, corrosion processes create huge problems of the world economy. The most conservative estimate of about 10% annually smelted metal is to replenish losses due to corrosion. However, the main corrosion damage not associated with loss of large amounts of metal, but with the failure of themselves metal products and structures as a result of loss of essential properties (strength, ductility, electrical conductivity, tightness, thermal conductivity, etc.). Protecting metals from corrosion - a global international problem, so in all developed countries attached great importance to anti-corrosion materials in all its manifestations. At present there five areas to combat the corrosion of metals:

1. Design, manufacture and application of corrosion-resistant materials for the manufacture of pipelines and process equipment
2. Create a corrosion-resistant coatings and methods and technologies for processing of material surfaces exposed to the corrosive effect
3. Creation and application of corrosion inhibitors

4. The use of electrochemical methods of protection of process equipment, pipelines and underground utilities in general
5. The package of measures for the development, design, construction and operation of pipelines and processing equipment in order to avoid the stress state of metal in which the corrosion processes are significantly faster.

Since the purpose of corrosion inhibitors do not exist, they are almost for each specific system, so the range of developed and produced by inhibitors of the world is enormous. In this world of science and technology known to more than 300000 names of individual chemicals and various technical formulations are classified as corrosion inhibitors. The volume of world production and consumption of corrosion inhibitors, lubricant additives is 4.4 million tons/year with a growth trend 5.5-6.0 million tons/year. Of this amount: the inhibition of oil, gas, produced water and other environments in oil and gas industry - 20-25%, the preparation and processing of oil - 2-5%, the inhibition of oil production and protection of oil-based, 65-75% for other needs (inhibition of acid cooling media, etc.) - 3-5%. A large number of inhibitors used in pipeline transportation.

Corrosion inhibitors are the most diverse classes of organic and inorganic compounds, most of which is synthetic, not naturally occurring. It is obvious that the negative impact on the biosphere biologically hostile, recalcitrant compounds can not be overestimated, especially in a wide range of use of a reagent. The problem of corrosion of equipment, in particular oil field, is successfully solved, but to date there is no product that simultaneously satisfies all the requirements of the requirements: high efficiency, low cost, versatility and environmental safety. In light of this ever being the development of new corrosion inhibitors, which would be advantageous to differ from the existing not only the efficiency of inhibition, but also environmental safety.

For example, a corrosion inhibitor designed on the basis of waste vegetable oil that contains no toxic compounds and are low cost at a degree of protection of 78-95% depending on the environment.

BASF's proposed "green" corrosion inhibitors based on propargyl alcohol, known under the trade name Korantin ® PM (Figure 6).

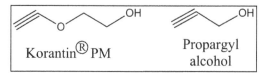

Fig. 6. BASF's corrosion inhibitors

Their most significant differences is not only cheap and nontoxic, but high inhibitory activity, 2-3 times exceeding some used products.

Company Cortec patented volatile corrosion inhibitor, a major component of which is ammonium benzoate - a non-toxic compound to the environment. The use of such compounds is justified in the manufacture of new modern packaging materials to prevent not only mechanical damage and corrosion of metal products during transportation and storage.

It is known that the salts of chromium (III) are used as components of an effective corrosion inhibitors of metal products, but their use should be limited due to toxicity. Creating a less risky alternative to trains - the use of titanium salts in combination with oxidants such as hydrogen peroxide and the use of mono-, oligo-and polysaccharides as film-forming component. Described as an inhibitor of the phosphoric and boric acids with much less toxicity.

The problem of corrosion protection equipment, operated in hostile environments such as seawater, is quite serious. Offshore drilling and production of oil and gas offshore require not only high but also low toxicity and biodegradable oil-reagent did not affect the indigenous inhabitants of the marine flora and fauna. Found that many of the known corrosion inhibitors, even at a concentration of less than 1 ppm inhibit the growth of algae. Preferably, the inhibitor is biodegradable at least 60% within 28 days after release to the environment. Also an important requirement is that sufficient hydrophilicity to minimize the bioaccumulation in adipose tissue, since lipophilic substances tend to accumulate as they move up the food chain.

Established that the imidazolines are not only effective inhibitors of corrosion, but also promising from an environmental point of view of satisfying the rigorous standards of toxicity. They do not contain phosphorus and sulfur atoms, and therefore are considered more "environmentally friendly". However, this does not mean that the substance is completely harmless, it is a very low toxicity at relevant concentrations.

One of the leading companies engaged in the development and manufacture of oilfield chemicals - Baker Huges proposed as corrosion inhibitors imidazolines (Figure 7)

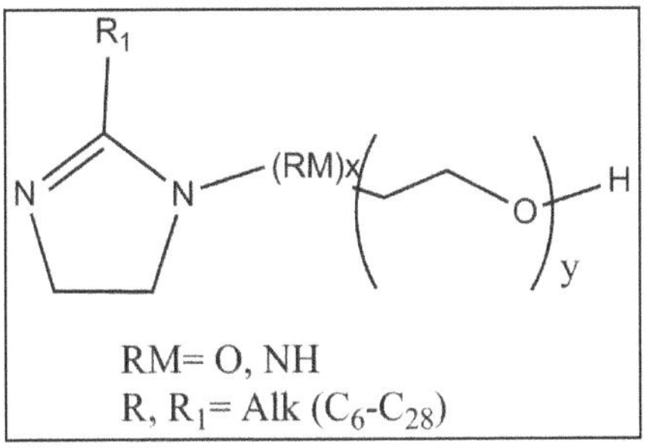

$$RM = O, NH$$
$$R, R_1 = Alk (C_6\text{-}C_{28})$$

Fig. 7. Imidazoline derivatives- corrosion inhibitors

In the product line, the Russian company also has Mirrico anticorrosive agents on the basis of imidazolines on brand Scimol.

To protect the equipment with high concentrations of hydrogen sulfide and carbon dioxide proposed a corrosion inhibitor, which is a cyclic lactone (Figure 8) disaccharide ester and fatty acid obtained by enzymatic synthesis using the culture of *Torulopsis apicola* relevant material: oligo-and polysaccharides, saturated / unsaturated fatty acids oils, fats and other substances of plant or animal origin.

Along with a significant inhibitory effect observed with the following positive aspects: easy dispersibility in water, good adsorption on metal surfaces for values of pH 5.3 or lower, non-inflammability and significantly less toxicity compared to known inhibitors.

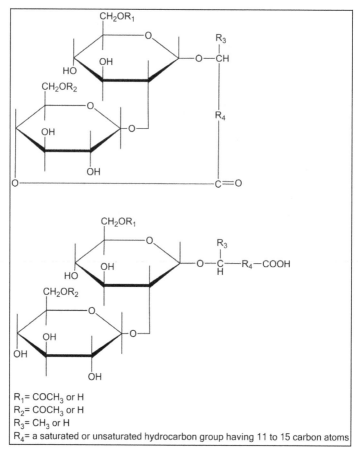

R$_1$= COCH$_3$ or H
R$_2$= COCH$_3$ or H
R$_3$= CH$_3$ or H
R$_4$= a saturated or unsaturated hydrocarbon group having 11 to 15 carbon atoms

Fig. 8. Corrosion inhibitors based on oligo-and polysaccharides

Despite the variety of substances used as corrosion inhibitors, the most promising in terms of 'green' chemistry, are substances of natural origin and their various derivatives.

One of the main features of oilfield chemicals is their variety, caused by various conditions of oil production, changing the composition of crude oil and its associated waters. It often happens that the composition of oil in the well is so unique that it has required the selection of specific agents with the optimal dose is for a particular hole. The problem of corrosion of oilfield equipment depends on the chemical composition of the medium in which it is located: it is saturated salt solutions and contain significant amounts of hydrogen sulfide and mercaptans, water-oil emulsion and produced water with dissolved carbon dioxide, so the selection and creation of appropriate reagents is often a nontrivial task.

5.3 Scale inhibitors

The present stage of oil production is characterized by the need to make the surface of the vast amounts of associated water reservoir as well as injected in the reservoir to maintain pressure.

As a result of flooding of the output at all stages of oil is the formation of salt deposits. Accumulating in the production strings on the surface of downhole pumping equipment and systems infield collection and preparation of oil, salt deposits not only cause great material costs in the process of removing them, but also to a significant loss in oil production.

The effectiveness of measures to combat the deposition of salt depends on an integrated approach to solving this problem. Need to know the physical and chemical processes and the causes of salt formation, taking into account the geological and physical conditions of occurrence of oil, human events and features of the development of oil and gas deposits and operating wells. The key areas for the deposition of salts from oil should be warning them, as a permanent measure, based on the best technological solutions that require scientific and methodological generalizations and systematic approach.

Based on the economic viability, depending on the conditions and characteristics of reservoir development, availability of raw materials, availability of technical resources and other factors may have different ways of dealing with this phenomenon, but in practice the problem of oil-field warning of salt deposits is decided mainly by inhibitor protection of wells and equipment .

Historically, the dominant classes of products inhibiting scaling in oil and gas production were phosphorous substances (eg aminomethylenephosphonates and phosphonic acids) and synthetic water soluble polymers such as polimaleates, polyacrylates, polisulfonates. Their main disadvantages are toxic to the environment and bioundegradability. The use of both classes of inhibitors are more tightly constrained as a result of legislative control. In recent years, the market introduced the latest "green" inhibitors: poly (amino acid) and chemically enhanced natural materials. This class of products has a low environmental toxicity, but the use of these inhibitors is still low because of the difficulty of their synthesis and significant economic costs compared with existing technologies.

Company AkzoNobel has been developed a new class of materials - hybrid polymers (copolymers of polysaccharides and polycarboxylic acids), (Figure 9) which combine the advantages of a single molecule, and synthetic and natural materials, the lack of benefit other than some of the restrictions currently known "green" products.

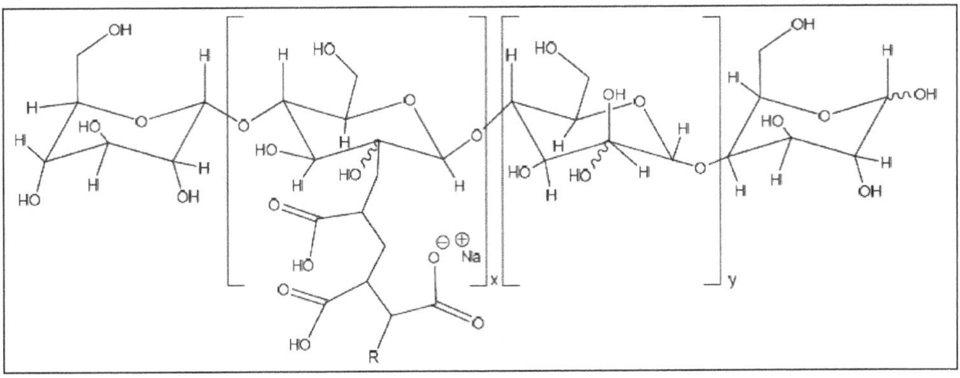

Fig. 9. AkzoNobel's hybrid polymers

Upon receipt of target compounds can be used a wide range of synthetic monomers, and the polysaccharide framework gives different functional characteristics while preserving the key properties of inhibitors. Studies conducted by the company revealed additional benefits of hybrid polymers, such as biodegradable and nontoxic. In addition, they are more environmentally sustainable solutions for the inhibition of salt deposition than synthetic products and not accumulated by living organisms.

Test (GLP standard)	Method	Result	Goal
Bioaccumulation	OECD 107	lgP_{ow}-2.95	lgP_{ow}<2.95
Acute toxicity - algae	ISO 10253	EC_{50} >100 mg/l	EC_{50} >100 mg/l
Acute toxicity - halibut fry	OECD 203	LC_{50} >100 mg/l	LC_{50} >100 mg/l
Acute toxicity - *acartia tonsa*	ISO TC147/SC5/WG2	LC_{50} >100 mg/l	LC_{50} >100 mg/l

Table 1. Biodegradability test results

These hybrid polymers are produced by more than 50% of renewable raw materials, as opposed to synthetic materials, and have lower emissions of carbon dioxide - hybridization reduces CO2 emissions by more than 50%.

The company also points to the adaptability of the technology of synthesis of target compounds, which expands the possibilities for optimizing the properties of the products according to specified criteria.

The company Rhodia offers to use as an inhibitor of salt deposits (carbonates and sulfates of magnesium, calcium and barium) natural polysaccharide composed of mannose and galactose residues in the native form and after modification of the relevant reagents (Figure 10). It is noteworthy that such products are not only patented, and commercially available.

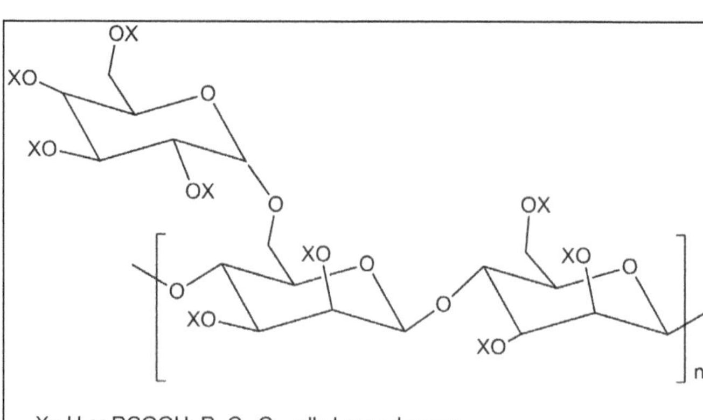

X= H or RCOOH; R=C_1-C_{22} alkyl or aryl group

Fig. 10. Rhodia's polymeric scale inhibitors

It is also noted that in the future as raw material for production may be other polysaccharides: starch and cellulose.

5.4 Biocides

Microorganisms that cause biological corrosion, play a significant role in the corrosion of underground oil, gas and water pipelines, corrosion of ship and aircraft equipment, metallurgy and metalworking equipment, chemical and food industries. Microorganisms act as corrosive agents mainly due to the aggressive production of metabolites and a corrosive environment. As an aggressive advocate of metabolites of organic and inorganic acids, enzymes, and hydrogen sulfide.

In many cases, the microbiological contamination is obvious, with visible signs include changes in color medium viscosity, presence of slime, sludge and other sediments. Less obvious indicators - the inefficiency of film-forming corrosion inhibitors and the presence of contaminants such as salts or iron oxides.

In the oil industry, the main agents of the corrosion process of metal fracture are sulfate-reducing bacteria. They account for at least 90% of hydrogen sulfide entering the cycle of sulfur compounds that about 80% of all corrosion damage land and 50% damage of underground equipment. The economic costs of biological damage can be up to 3% of the metal structures are exploited.

Spectrum inhibitors, bactericides and fungicides used to suppress the activity of microorganisms is extremely varied, but among them are many toxicants not only to microorganisms but also to warm-blooded. For example found that benzotriazole is used as a bactericide has high toxicity to mammals, and some arthropods.

Among the aldehydes used in the manufacture of disinfectants found formaldehyde, glutaric and orthophthalic aldehydes having a broad spectrum of activity (Gram-positive and Gram-negative bacteria, fungi, mycobacteria, shell / implant failure viruses), including spores. Drugs, having in its composition glutaraldehyde get better, biocidal properties do not cause corrosion of materials of tools will not damage fabrics and surfaces are stable (which allows multiple solutions), have good penetration, fast destructible in wastewater. In fact, disinfectants on the basis of glutaraldehyde has been and remains the "gold standard" in many spheres of human activity. We can say that among the classes of oilfield chemicals biocides are the most "green" reagents.

5.5 New types of oilfield chemicals

5.5.1 Reagents complex action

Reagents complex action can perform several functions simultaneously, for example, act as a corrosion inhibitor and scale inhibitor. In this case, it is not a simple mixture of several substances, and about a component that performs both functions. As often occurs in oil fields, several adverse events (corrode equipment, salt deposition, the presence of sulfate-reducing bacteria, etc.), the creation of such reagents can significantly reduce the costs of minimizing damage.

An example of a complex of the reagent is a corrosion inhibitor and scaling based on aspartic acid, proposed by Nalco. Synthesis is proposed to conduct a phase 2 (Figure 11).

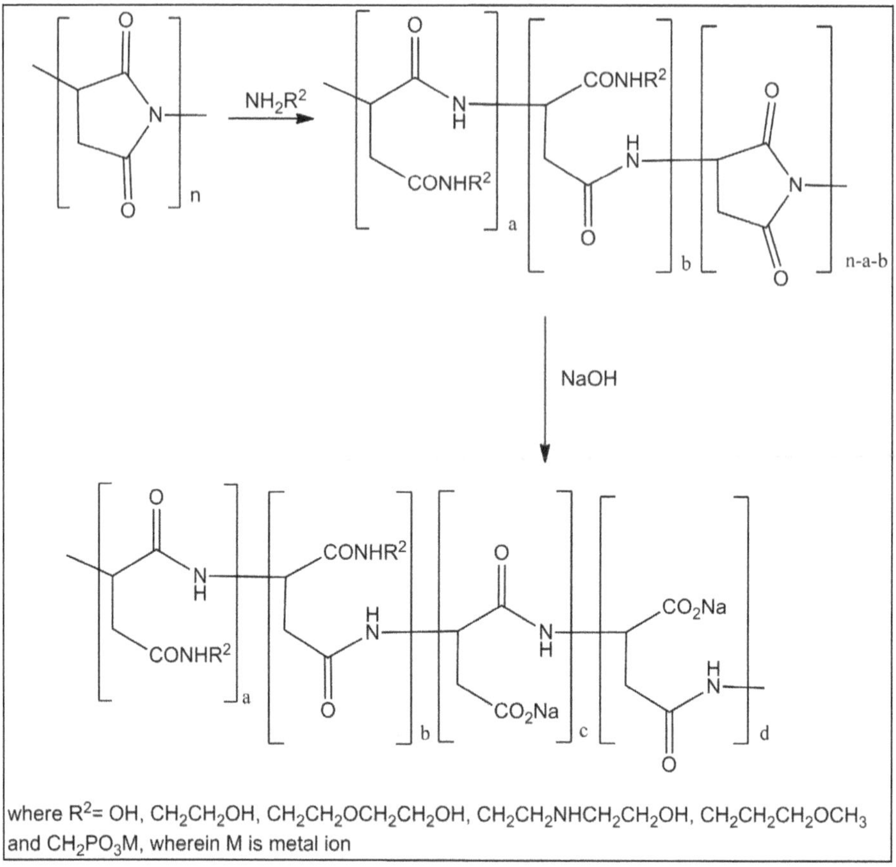

where R^2= OH, CH_2CH_2OH, $CH_2CH_2OCH_2CH_2OH$, $CH_2CH_2NHCH_2CH_2OH$, $CH_2CH_2CH_2OCH_3$ and CH_2PO_3M, wherein M is metal ion

Fig. 11. Synthesis of the reagent complex action

Highlights not only the protective properties, and biodegradability of these products with polymeric structure.

5.5.2 Encapsulated reagents

Along with the creation of new reagents for oil production, become important technological methods of their use. Depending on the reagents can be used for the following technologies:

- by continuous or periodic supply to the system using metering devices;
- periodic injection of a solution into the well followed by his supply in bottomhole zone.

Consistently can be used combined methods of delivery reagent, for example, initially periodic injection, and then - after 2-6 months - a continuous or periodic supply dosage of reagent in annulus.

Distribution is the method of periodic filing solution in annulus, but it is not efficient, since the low dynamic poles reagent quickly carried away by fluid flow. In the most favorable conditions of high dynamic frequency of feeding poles inhibitor is 15-20 days.

The method of dispensing reagents applied for maintenance of underground equipment and pipe elevator, but when there are problems in the bottomhole formation zone is necessary to supply into the reservoir. Dosing of reagent into the well (the system) is considered a reliable method, although it requires constant monitoring and maintenance of metering pumps and devices.

Disadvantage of the periodic injection of the solution in the annulus is that the process is accompanied by illuviation of reagent into the rock. A common shortcoming of these methods is too high consumption of reagent due to inefficiency of its use.

The use of submersible containers dispensers - the most economical and efficient way compared to others. The operating principle of these containers is based on different processes: dissolution, or leakage to the gravitational turbulent mixing of the reagent with the formation fluid. Submerged containers deliver reagent deep wells, where a minimal amount of active ingredient. However, the rate of dosing is determined by the downhole conditions, so the design of submerged container should be selected individually for each well. Usually this is not done or can not do so while the container is often different from the statements to the same dosing containers require special arrangements for their installation and maintenance.

It is now one of the most promising directions is the creation of microscopic containers, hoppers, the so-called microencapsulation reagent.

The advantages of using microencapsulated chemicals include:

- the possibility of long-acting agents required, as a consequence of the increase interservice interval;
- the possibility of joint use of substances of different classes that can react with each other for direct use;
- a longer residual effect of chemical treatment;
- a safer treatment with chemical agents;
- more simple equipment needed to handle well, the lower price, due to better control and less consumption of reagents;

The disadvantages of using such technology products are:

- tendency to adsorb on the capsule surface solids;
- reducing the duration of the reagent in the wells with high flow rates of fluids:

Microencapsulated product is a reagent, a prisoner in the polymer capsule. The reagent is pumped into the annulus, after which the capsules are deposited on the face. When operating the wells polymer membrane is dissolved, mixed with a liquid reservoir, providing a way out of the reagent.

Within 1-2 days after the product is well increased yield of reagent and then the system comes into equilibrium and its removal is lower (but above the minimum inhibitory concentration) and permanent.

Exit the reagent is carried out by the concentration gradient. With increasing consumption of reagent in the external environment is intensifying its removal from the capsule, and vice versa when - flow of reagent bounces back - his removal from the capsule slows down.

Material for the microcapsules are hydrophilic colloidal materials:

- protein origin (gelatin, albumin);
- polysaccharide structure (alginates, such as sodium alginate, casein, agar-agar, starch, pectin, carboxymethylcellulose, and some other). To obtain such materials are usually used to form a microcapsule by coacervation. The preferred material for encapsulation is gelatin, which precipitated a salt such as sodium sulfate or ammonium sulfate, or involved in coacervation with polyanions, such as gum arabic (acacia), even more preferred is the use of compounds capable of crosslinking of the carrier material, such as formaldehyde or glutaraldehyde (Figure 12).

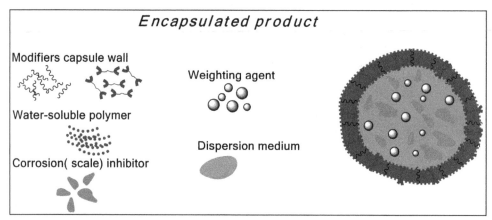

Fig. 12. Composition of the capsule

Encapsulated reagents offer oilfield companies such as Champion Technologies and Mirrico.

6. Conclusion

Thus, we can formulate the basic requirements for the product oil and gas industry in line with the standards of "green chemistry":

- High efficiency
- Availability
- Universality
- Environmental safety
- A sufficiently high rate of biodegradation into harmless substances
- Low cost.

However, the development and introduction of such products can be traced obvious difficulties.

- To develop and test new recipes, rather strictly limited requirements of "green chemistry" can take quite a long time because it is theoretically and practically more Bole complex than simply the development of new agents without at least their biological hazards.
- Replacement of existing products with new chemical industry requires, as a rule, a significant restructuring of both the technological process and equipment used, which involve significant economic costs.
- Financial investment in scientific research to develop "green" products industry can be quite large, so you need constant monitoring and adjustment of balance between environmental safety, high effectiveness and cost of the desired product.

In recent years, clearly observed trends of developed countries identified the use of products "green chemistry" on the basis not only for reasons of environmental safety, development of similar products and technological schemes involves minimal risk to humans and nature, a more rational approach to the concept of obtaining a product, which ultimately account may lead to a significant reduction in price. Of course, this ideal case, but the examples in the modern chemical industry there that points to the undoubted promise of modern science and technology for "green" way, and it's not just a fad and a means of promoting new products, and a new stage of development of chemical technology, industry and science.

7. Acknowledgment

Authors would like to acknowledge to top managers of GC «Mirrico»: Malyhin I. and Ramazanov R.

8. References

Anastas, P. & Warner, J. (1998). *Green Chemistry: Theory and Practice*, Oxford University Press, ISBN 0-198-50234-6, New York

Ash, M. & Ash, I. (2004). *Handbook of preservatives*, Synapse Information Resources, ISBN-10 1890595667, Endicolt

Becker, J.(1998). *Corrosion and scale handbook*, PennWell Books, ISBN 0878147497, Tulsa

Durkee, J. (2006). *Management of industrial cleaning technology and processes*, Elsevier, ISBN-10 0080448887 Oxford

Lunin, V., Tundo, P. & Lokteva, E. (2005). *Green Chemistry in Russia*, Poligrafica Venezia, ISBN 5-211-05001-0, Venezia

Sastri, V. (1998) *Corrosion inhibitors: principles and applications*, Willey, ISBN 0-471- 7608, New York

Schramm, L. (2000) *Surfactants: fundamentals and applications in the petroleum industry*, Cambridge University Press, ISBN 0-521-64067-9 Cambridge.

Karsa, D., & Ashworth, R. (2002). *Industrial biocides: selection and application*, Royal Society of Chemistry, ISBN 0-854-04805-7, Cambridge

Scott, E., & Gorman, S. (2001). *Glutaraldehyde. Disinfection, sterilization and preservation,* , Lippincott Williams&Wilkins, ISBN 1-4051-0199-7, New-York.

Cook, S. (1999). Green chemistry – evolution or revolution?. *Green Chemistry*, Oct. 1999, G138-G140.

Karabinos, J. & Ballum, A. (1954). tall oil studies. II. Decolorization of polyethenoxy tallates with ozone and hydrogen peroxide *J. Am. Oil Chem. Soc.*, Vol. 31 No 2, February 1954, pp. 71-74.

7

Greener Solvent-Free Reactions on ZnO

Mona Hosseini-Sarvari
Department of Chemistry, Faculty of Science, Shiraz University, Shiraz,
I. R. Iran

1. Introduction

Due to the growing concern for the influence of the organic solvent on the environment as well as on human body, organic reactions without use of conventional organic solvents have attracted the attention of synthetic organic chemists. Although a number of modern solvents, such as fluorous media, ionic liquids and water have been extensively studied recently, not using a solvent at all is definitely the best option. Development of solvent-free organic reactions is thus gaining prominence.

During the last decades, a central objective in synthetic organic chemistry has been to develop greener and more economically competitive processes for the efficient synthesis of biologically active compounds with potential application in the pharmaceutical or agrochemical industries. In this context, the solventless approach is simple with amazing versatility. It reduces the use of organic solvents and minimizes the formation of other waste. The reactions occur under mild conditions and usually require easier workup procedures and simpler equipment. Moreover, it may allow access to compounds that require harsh reaction conditions under traditional approaches or when the yields are too low to be of practical convenience. Because of economy and pollution, solvent-free reactions are of great interest in order to modernize classical procedures making them more clean, safe and easy to perform. Reactions on solid mineral supports, reactions without any solvent/support or catalyst, and solid-liquid phase transfer catalysis can be thus employed with noticeable increases in reactivity and selectivity. Therefore the following benefits could be mentioned for solvent-free conditions:

1. Avoid of large volumes of solvent reduces emission and needs for distillation.
2. Simple work-up, by extraction or distillation.
3. The absence of solvents facilitates scale-up.
4. Reactions are often cleaner, faster, and higher yielding.
5. Recyclable solid supports can be used instead of polluting mineral acids.
6. Safety is enhanced by reducing risks of overpressure and explosions.

Zinc oxide was known for a long time, since it was a by-product of copper smelting. ZnO has been intensively studied since 1935 (Bunn, 1935) and the theory of semiconductors got a firm start. The interest in ZnO is fueled and fanned by its direct wide band gap ($E_g \sim 3.3$ eV at 300 K), (Kakiuchi et al., 2006). ZnO also has much simpler crystal-growth technology, resulting in a potentially lower cost for ZnO base devices. ZnO has wide applications in the field of optoelectronics, (Kong & Wang, 2004) spintronics, (Sharma et al., 2003) piezoelectric transducers, (Catti et al., 2003) and ultraviolet optoelectronics, (Wang et al., 2004). Consequently, ZnO is widely used as an additive into numerous materials and products including plastics, glass, rubber, ceramics, lubricants, paints, ointments, pigments, etc. In addition, with the development of industrialization, organic chemists have been confronted with a new challenge of finding novel methods in organic synthesis that can reduce and finally eliminate the impact of volatile organic solvents and hazardous toxic chemicals on the environment. So, use of non-toxic, environmentally friendly, and inexpensive solid catalysts to perform organic reactions has attracted considerable interest. Due to this, great efforts have been made by different research groups to achieve the goal of making the ZnO as catalyst in organic transformations. Interesting results have been achieved, such as the use of catalytic amounts of ZnO alone or mixed metals and metal oxides with ZnO as catalysts. In fact, ZnO as a heterogeneous catalysts can be easily separated from the reaction mixture and reused; it is generally not corrosive and do not produce problematic side products. Different classes of organic transformations have been studied and utilized using ZnO as heterogeneous catalysts, mainly exploited in the production of fine chemicals, is the subject of intensive studies in this area. Many catalytic ability of ZnO have been explored for various organic reactions facilitating synthesis of fine and specially chemicals. The application of ZnO catalyst not only is industrially important but also has academic merit. The present chapter review deal with the development in the use of ZnO as catalyst for a diverse range of organic transformations in solvent free conditions. Solvent-free organic reactions envisaged in the literature by utilizing ZnO catalyst are organized and are outline as below.

2. Solvent-free reactions catalyzed by ZnO

2.1 Friedel-Crafts acylation

Aromatic ketones are valuable intermediates in the production of various fine chemicals, which are synthesized mainly by Friedel-Crafts acylation of aromatics with acid chlorides or carboxylic anhydrides (Jackson & Hargreaves, 2009). Traditionally, these reactions have been carried out using stoichiometric amounts of liquid Bronsted acids or Lewis acids. Also aromatic ketones are the valuable intermediates or final compounds used in the production of pharmaceuticals, cosmetics, agrochemicals, dyes, and specialty chemicals. Nowadays, the restrictions imposed by the waste-minimization laws and economic considerations driven to the development of new catalytic technologies. Modern processes are in fact, based on solid catalysts.

Different researcher groups have been reported that ZnO exhibit one of the best performance in the Friedel-Crafts acylation of various benzene derivatives in solvent free-conditions, (Ashoka et al., 2010; Hosseini-Sarvari & Sharghi, 2004; Thakuria et al., 2007; Wang et al., 2008), (Scheme 1).

$$\text{RCHO} + \text{Ar-H} \xrightarrow[\text{rt, solvent-free}]{\text{ZnO}} \text{ArCOR} + \text{HCl}$$

Scheme 1. Friedel-Crafts acylation on ZnO

2.2 Protection reactions

The use of protecting groups is very important in organic synthesis, often being the key for the success of many synthetic enterprises. The acylation of alcohols, phenols, amines, and thiols are an important transformation in organic synthesis, (Greene & Wuts, 1999). Acylation of such functional groups is often necessary during the course of various transformations in a synthetic sequence, especially in the construction of poly functional molecules such as nucleosides, carbohydrates, steroids and natural products.

Hosseini-Sarvari and Sharghi, reported the acylation reaction of alcohols, phenols and amines under solvent-free conditions employing ZnO as catalyst (Scheme 2), (Hosseini-Sarvari & Sharghi, 2005).

$$RXH \xrightarrow[\text{ZnO, rt, solvent-free}]{\text{RCOCl or } (R'CO)_2O} RXCOR'$$

R = alkyl and aryl; R' = Ph, Me; X=O, NH

Scheme 2. Protection of alcohols, phenols, and amines by ZnO

The catalyst was successfully applied for acylation of a diverse range of alcohols, phenols and amines. In the case of alcohols and phenols, an acid chloride was preferred over the corresponding acidic anhydride. The reaction with acid anhydride was too slow to have practical application. Both primary and secondary alcohols react very well and tertiary alcohol is also acylated smoothly without any side products observed. Also, the reactions of amines with Ac₂O were so fast in comparison to those of the aliphatic alcohols that the selective protection of an amine in the presence of aliphatic alcohols appeared to be a distinct possibility. Also, the amino group in aminophenol was selectivity acylated (Scheme 3).

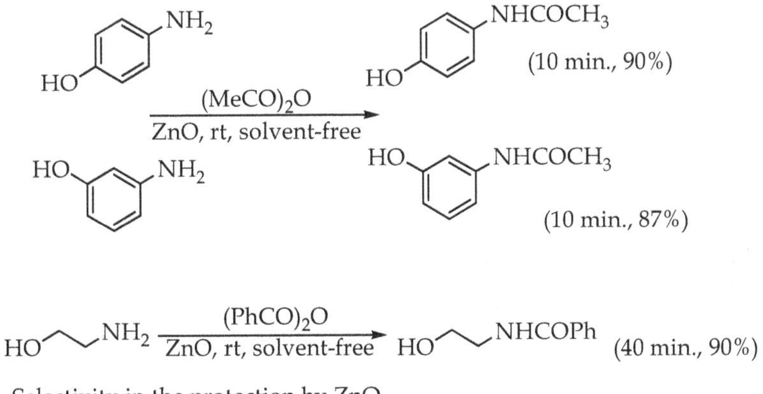

Scheme 3. Selectivity in the protection by ZnO

Similarly, in another studies, O-acylation of alcohols and phenols with acid chlorides were performed employing ZnO and nano ZnO, (Moghaddam & Saeidian 2007; Tammaddon et al., 2005; Tayebee et al., 2010).

Recently, Bandgar and co-workers reported a convenient and efficient synthesis of thiol esters via the reaction of acyl chlorides with thiols using ZnO as catalyst under solvent-free conditions at room temperature. Major advantages associate with this protocol includes mild reaction conditions, short reaction time, excellent yields and recyclability of the catalyst, (Scheme 4) (Bandgar et al., 2009).

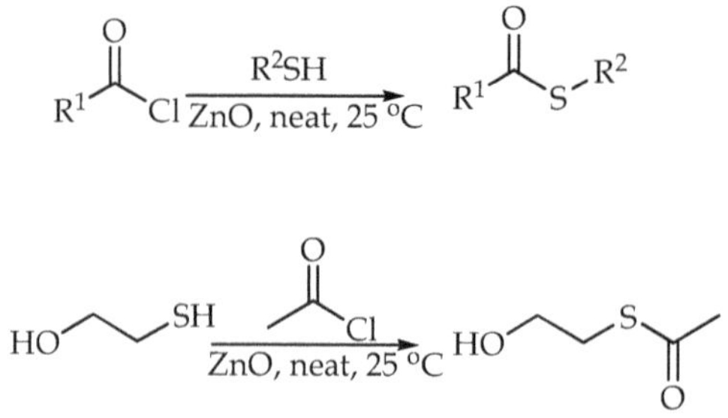

R^1, R^2 = -Ph	94%
R^1 = -t-Bu R^2 = -Et	76%
R^1 = -t-Bu R^2 = -Ph	96%
R^1 = -Me R^2 = -Ph	97%
R^1 = -Ph R^2 = -Et	89%
R^1 = -p-ClC$_6$H$_4$ R^2 = -Et	90%
R^1 = -p-OMeC$_6$H$_4$ R^2 = -Et	90%
R^1 = -CH$_2$Ph R^2 = -Et	90%
R^1 = -p-NO$_2$C$_6$H$_4$ R^2 = -Ph	80%

Scheme 4. Synthesis of thiol esters using zinc oxide

In addition, ZnO as economical and heterogeneous catalyst was reported for the silylation of alcohols, phenols and naphthols (Shaterian & Ghashang, 2007) (Scheme 5).

$$ROH + (Me_3Si)_2NH \xrightarrow[\text{Solvent-free, rt}]{\text{ZnO (cat.)}} R\text{-OTMS}$$

R= Aryl, primary, secondary, and tertiary aliphatic

Scheme 5. Silylation of alcohols, phenols, and naphthols using ZnO

2.3 Knoevenagel condensations

Knoevenagel (Knoevenagel, 1898) condensations which are one of the most important C-C bond forming reactions have been widely used in the synthesis of important intermediates or products for coumarin derivatives, cosmetics, perfumes, pharmaceuticals, calcium antagonists, and polymers.

Hosseini-Sarvari and co-workers (Hosseini-Sarvari et al., 2008) were reported that nano powder ZnO could be used as catalyst for *Knoevenagel* condensation. This condensation was performed using various aliphatic, aromatic, and heterocyclic aldehydes with malononitrile under solvent-free conditions in a one-step process. Most of the reactions investigated with nano ZnO catalysts were almost complete in 5 min to 3 h duration to produce the corresponding electrophilic alkenes in 90-98% yield (Scheme 6, and Table 1).

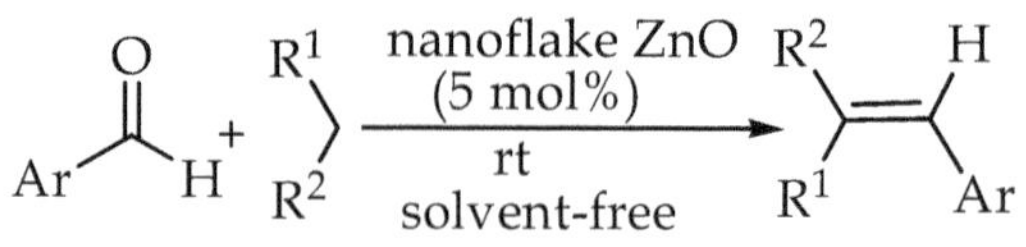

Scheme 6. Knovenagel condensation using nano ZnO

Entry	Ar	R^1	R^2	Time (min)	Isolated Yields (%)
1	Ph	CN	CN	210	90
2	$4\text{-MeC}_6\text{H}_4$	CN	CN	180	90
3	$4\text{-MeOC}_6\text{H}_4$	CN	CN	150	90
4	$2\text{-MeOC}_6\text{H}_4$	CN	CN	180	95
5	$4\text{-ClC}_6\text{H}_4$	CN	CN	10	98
6	$4\text{-HOC}_6\text{H}_4$	CN	CN	180	90
7	$4\text{-NO}_2\text{C}_6\text{H}_4$	CN	CN	10	98
8	$3\text{-ClC}_6\text{H}_4$	CN	CN	180	90
9	$2\text{-ClC}_6\text{H}_4$	CN	CN	30	90
10	2-Thienyl	CN	CN	120	90
11	4-Pyridyl	CN	CN	5	98
12	2- Furyl	CN	CN	180	95
13	$4\text{-ClC}_6\text{H}_4$	CN	CO_2Et	240	90
14	$4\text{-ClC}_6\text{H}_4$	CN	OMe	1440	0
15	$4\text{-ClC}_6\text{H}_4$	CO_2Et	Cl	60	90
16	$4\text{-ClC}_6\text{H}_4$	CO_2Et	CO_2Et	300	90

Table 1. Knoevenagel condensation using nano flake ZnO (0.004 g) at 25 °C under solvent-free conditions

2.4 Biginelli reaction

The three-component condensation reaction between aldehyde, β–ketoester, and urea under strong acidic conditions to furnish 3,4-dihydropyrimidin-2(1H) was known as the Biginelli reaction.

Bahrami and co-workers reported a simple, efficient and practical procedure for the Biginelli reaction using ZnO as a novel and reusable catalyst under solvent-free conditions (Bahrami et al., 2009). The reaction proceeds efficiently under these conditions, and the dihydropyrimidiones were produced in high yields (Scheme 7).

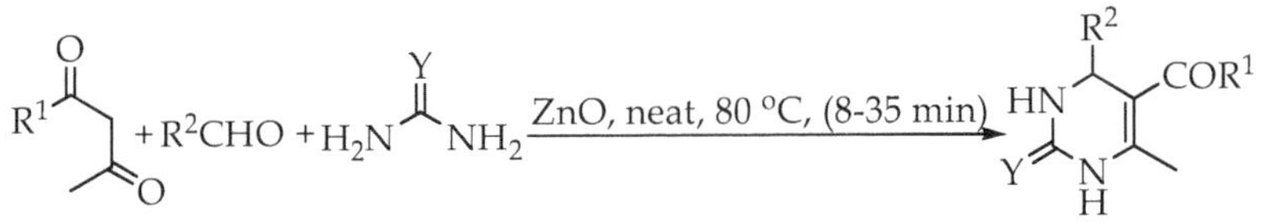

Scheme 7. Solvent-free synthesis of dihydro pyrimidinones catalyzed by ZnO

2.5 Hantzsch condensation

Described more than one century ago by Hantzsch, (Hantzsch, 1881) dialkyl 1,4-dihydro-2,6-dimethylpyridine-3,5-dicarboxylates (1,4-DHP) have now been recognized as vital drugs. 1,4-DHP derivatives possess a variety of biological activities such as vasodilator, bronchodilator, anti-atherosclerotic, anti-tumor, geroprotective, hepatoprotective and anti-diabetic activity. In addition, recently preceding studies have suggested that 1,4-DHP derivatives also provide an antioxidant protective effect that may contribute to their pharmacological activities. Oxidation of 1,4-DHP to pyridines has also been extensively studied. These examples clearly indicate the remarkable potential of novel 1,4-DHP and polyhydroquinoline derivatives as a source of valuable drug candidates and useful intermediates in organic chemistry. Many homogeneous and heterogeneous catalysts have been reported for the preparation of 1,4-DHP *via* the Hantzsch condensation, (Heravi et al., 2007; Kumar & Mauria, 2007).

Hantzsch condensation was thoroughly investigated employing ZnO catalyst, (Katkar et al., 2010, 2011; Moghaddam et al., 2009). Kassaee and co-workers employed ZnO nano particles as an efficient and heterogeneous catalyst for synthesis of polyhydroquinoline derivatives under solvent free conditions at room temperature (Kassaee et al., 2010). They were shown that in comparison with the same reaction catalyzed by commercially bulk ZnO (Moghaddam et al., 2009), use of ZnO nano particles reduced the reaction time with higher yields (Scheme 8).

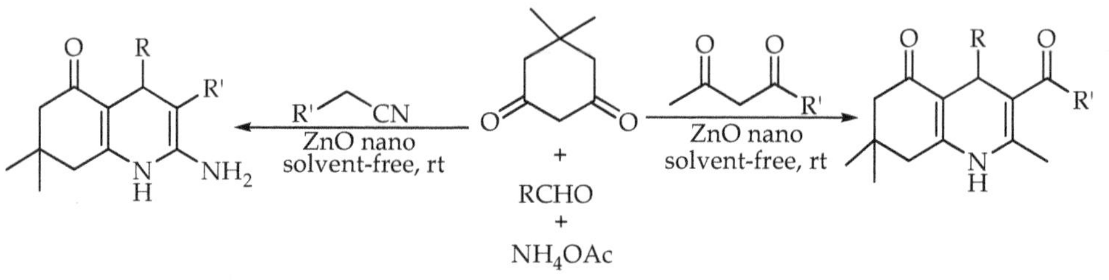

Scheme 8. Hantsch condensation catalyzed by nano ZnO

The catalytic activity and the ability to recycle and reuse ZnO nano particles were studied in this system (Table 2). The catalyst was separated by centrifuging the aqueous layer at 3,000 rpm at 20 °C for 3 min, and was reused as such for subsequent experiments under similar reaction conditions.

Run	Yields (%)	Recovery of nano ZnO (%)
1	98	99
2	97	99
3	97	98
4	93	96

Table 2. Reusability of the ZnO nano particles catalyst

2.6 Phospha-Michael addition

Similar to the Michaelis-Arbuzov and the Michaelis-Becker reaction the phospha-Michael addition, *i.e.* the addition of a phosphorus nucleophile to an acceptor-substituted alkene or alkyne, certainly represents one of the most versatile and powerful tools for the formation of P-C bonds since many different electrophiles and P nucleophiles can be combined with each other. This offers the possibility to access many diversely functionalized products. This reaction was investigated employing nano ZnO catalyst. Hosseini-Sarvari et al. reported that nano flake ZnO exhibits the best performance in the phospha-Michael addition of phosphorus nucleophile to α,β–unsaturated malonates under solvent-free conditions at 50 °C (Scheme 9) (Hosseini-Sarvari & Etemad, 2008).

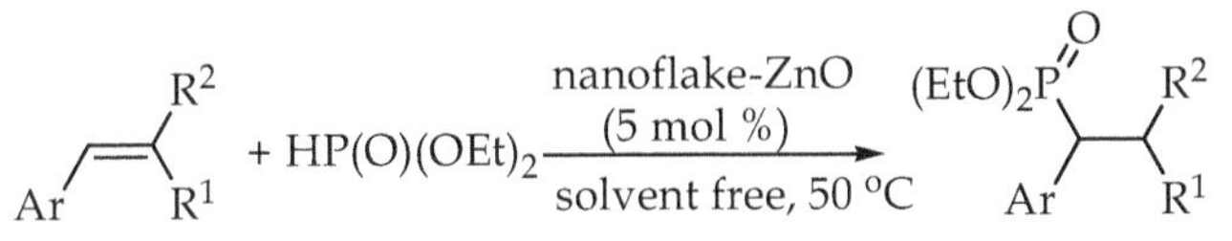

R^1 = CN; R^2 = CN, CO_2Et

Scheme 9. Phospha-Michael addition over nano ZnO

The authors were shown that, two kinds of ZnO (commercial ZnO and nano flake prepared ZnO, (20-30 nm)) were screened in the reaction between diethyphosphite and 2-((4-chlorophenyl)methylene)malononitrile (Table 3). As shown in Table 3, nano flake ZnO was found to be more effective than commercially ZnO in mediating the phospha Michael addition under solvent-free conditions. In order to examine the solvent effect and in quest for the deployment of a benign reaction medium, the reaction was explored in CH_2Cl_2, CH_3CN, THF and water. The reaction in solvents required relatively longer reaction times and afforded moderate yields of the product.

Entry	Catalyst	Solvent	Time (min)	Isolated Yield (%)
1	Commercially ZnO	Non	120	80
2	Nano flake ZnO	Non	30	98
3	Nano flake ZnO	CH_2Cl_2	180	43
4	Nano flake ZnO	CH_3CN	180	40
5	Nano flake ZnO	THF	180	40
6	Nano flake ZnO	H_2O	180	0

Table 3. Reaction between 2-((4-chlorophenyl)methylene)malononitrile with diethyl phosphite catalyzed by different crystallite of ZnO at 50 °C

2.7 N-Formylation of amines

Formamides are a class of important intermediates in organic synthesis. They have been widely used in the synthesis of pharmaceutically important compounds. A numerous methods have been reported for the formation of formamides (Green & Wuts, 1999). However, there are several factors such as low yield, difficulties in workup procedure and use of expensive reagents limiting their applications. This transformation was thoroughly investigated employing ZnO catalyst.

Recently, ZnO under solvent-free conditions have proved to be useful and reusable catalyst for N-formylation of amines using aqueous formic acid (85%) as formylating agent. This reaction was performed using various aliphatic, aromatic, heterocyclic primary and secondary amines under solvent-free conditions (Scheme 10) (Hosseini-Sarvari & Sharghi, 2006).

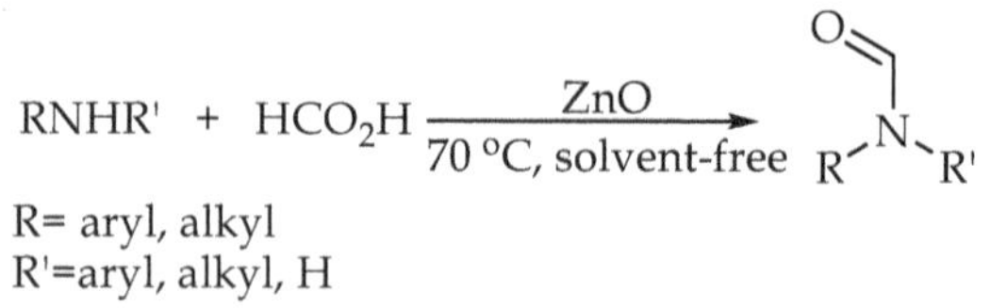

$$\text{RNHR'} \ + \ \text{HCO}_2\text{H} \xrightarrow[\text{70 °C, solvent-free}]{\text{ZnO}} \quad \underset{R \diagdown N \diagup R'}{\overset{O}{\diagup\!\!\!\!\diagdown}}$$

R= aryl, alkyl
R'=aryl, alkyl, H

Scheme 10. N-Formylation of amines over ZnO

Also, prepared macroporous ZnO in presence of agar gel as template, later heated to 600 °C to produce the ZnO, has been used for the N-formylation of aniline (1 mmol) with formic acid (2.5 mmol) in solvent free conditions as indicated in Table 4 (Thakuria et al., 2007).

Entry	Catalyst	Catalyst (mmol)	Time (min)	Isolated Yield (%)
1	Commercial ZnO	0.50	10	99
2	ZnO (macroporous)	0.50	08	99
3	ZnO (macroporous)	0.25	120	55

Table 4. Comparison of N-formylation of aniline (1 mmol) with formic acid (2.5 mmol) by using ZnO as catalyst

2.8 N-Alkylation of imidazoles

Imidazole-4-carboxaldehyde and 4-cyanoimidazole were N-benzylated and N-methylated using benzyl chloride and methyl iodide on ZnO under basic conditions without solvent (Oresmaa et al., 2007). Oresmaa et al. reported that Et₃N or K₂CO₃ was added as base in the reaction on ZnO. They also investigated the effect of bases and catalyst on the product distribution of 1,4- and 1,5- substituted compounds (Scheme 11).

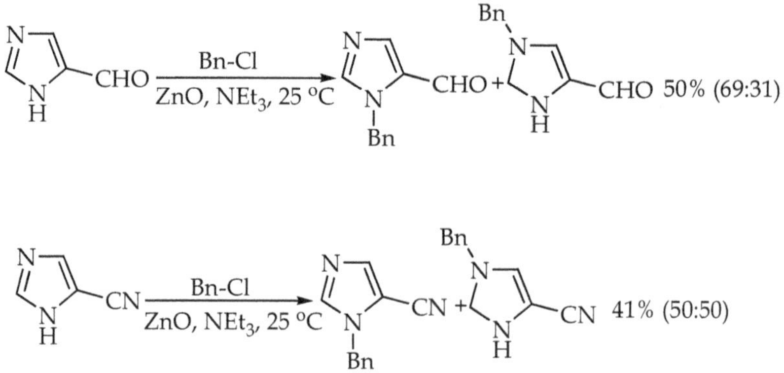

Scheme 11. N-Alkylation of imidazoles over ZnO

2.9 Synthesis of oxazolines

Oxazolines have been of great interest due to their versatility as protecting groups, as chiral auxiliaries in asymmetric synthesis, and as ligands for asymmetric catalysis. ZnO was a useful catalyst for the preparation of 2-oxazolines from carboxylic acids (Dotani, 1997; Ishikawa, 1999; Morimoto & Ishikawa, 1997).

Recently, Garcia-tellado and co-workers (Garcia-tellado et al., 2003) investigated the synthesis of 4, 4-disubstituted-2-oxazoline using ZnO in solvent-free microwave-assisted conditions. They were shown that zinc oxide was found to be the most efficient. It acts both as a solid support and as a soft Lewis acid catalyst. Other acidic solid supports (montmorillonites KSF and K10, calcinated Al_2O_3, SiO_2) were unsuccessful. They also described that the zinc oxide seems to play a double role: it creates a polar environment for the microwave catalysis (polar solid support) and activates the carbonyl group for the condensation (Lewis acid catalyst) (Scheme 12).

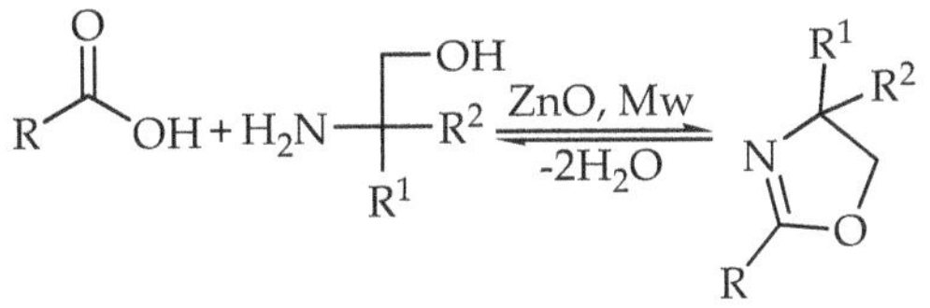

Scheme 12. Synthesis of 4, 4-disubstituted-2-oxazoline using ZnO

2.10 Synthesis of benzimidazole derivatives

Recently, a synthesized nano particle ZnO catalyzes the synthesis of benzimidazoles with formic acid in excellent yields (Scheme 13) (Alinezhad et al., 2012). Benzimidazoles are important natural and synthetic heterocyclic compounds. Some of their derivatives are marketed as antifungal, antihelmintic, and antipsychotic drugs and other derivatives have been found to possess some interesting bioactivities. The method which was reported by Alinezhad and co-workers avoids the use of expensive reagents and the reaction is performed under solvent-free condition, making it efficient and environmentally benign. The advantages of this method include the ease of preparation of nano particle ZnO; reusable, nontoxic, and inexpensive heterogeneous nano catalyst; mild reaction conditions; easy and clean workup; and convenient procedure.

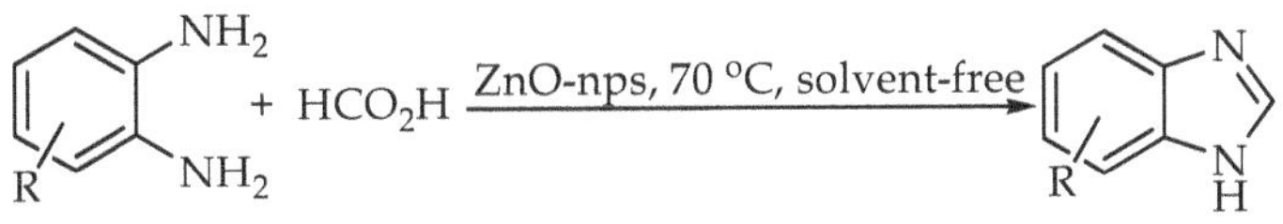

Scheme 13. Synthesis of benzimidazoles by nano particle ZnO

2.11 Synthesis of quinolines

Quinolines are well known for a wide range of medicinal properties being used as antimalarial, antiasthmatic, antihypertensive, antibacterial and tyrosine kinase inhibiting agent. Recently, the author reported syntheses wide range of quinolines have investigated by using nano ZnO as a heterogeneous solid catalyst under solvent-free condition (Scheme 14) (Hosseini-Sarvari, 2011).

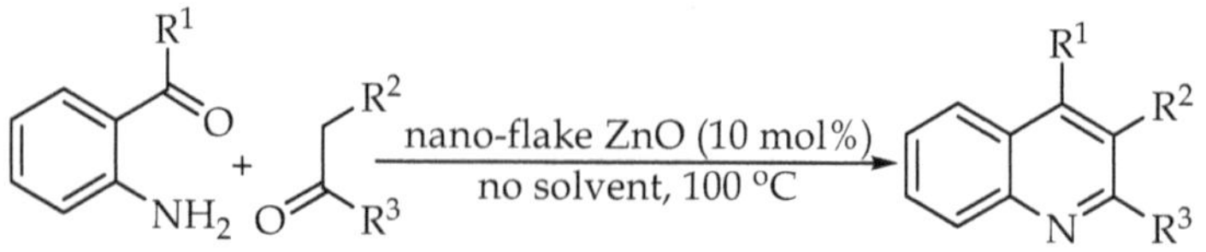

Scheme 14. Synthesis of quinolines catalyzed by nano ZnO

From Table 5, the author examined some other metal oxides in the synthesis of ethyl 2,4-dimethylquinoline-3-carboxylate. Also, in order to examine the solvent effect, the reaction was explored in toluene, THF, CH_3CN, and water. The reaction in organic solvents required relatively longer reaction times and afforded trace yields of the product.

Entry	Catalyst (mol %)	Solvent	Time (h)	Isolated Yield (%)
1	Nano-flake ZnO (10)	None	4	98
2	Nano-flake ZnO (10)	Toluene	24	trace
3	Nano-flake ZnO (10)	THF	24	trace
4	Nano-flake ZnO (10)	CH_3CN	24	trace
5	Nano-flake ZnO (10)	H_2O	24	0
6	Nano-flake ZnO (5)	None	12	85
7	Nano-flake ZnO (2)	None	24	67
8	Nano-flake ZnO (20)	None	4	95
9	Nano-particle ZnO (10)	None	12	80
10	Commercially ZnO (10)	None	11	90
11	Commercially MgO (10)	None	24	40
12	Commercially TiO_2 (10)	None	8	75
13	Commercially CaO (10)	None	24	73

Table 5. Synthesis of ethyl 2,4-dimethylquinoline-3-carboxylate at 100 °C

2.12 Synthesis of nitriles

The cyano moiety is a highly important not only due to its synthetic value as precursor to other functionalities but also due to its presence in a variety of natural products, pharmaceuticals and novel materials. Although a plethora of methods are known for access to the cyano functionality, dehydration of aldoximes remains a convenient route. ZnO has been used for the synthesis of Nitriles from the dehydration of aldehydes under solvent-free conditions (Hosseini-Sarvari, 2005; Reddy & Pasha, 2010). The author reported the combination of ZnO and acetyl chloride accelerated the catalytic dehydration of aldoximes into nitriles dramatically (Scheme 15).

$$\mathrm{R(Ar)CHO} \xrightarrow[\mathrm{NH_2OH.HCl,}\ 60\ ^{\circ}\mathrm{C}]{\mathrm{ZnO}} \mathrm{R(Ar)CH{=}NOH} \xrightarrow[80\ ^{\circ}\mathrm{C}]{\mathrm{ZnO}} \mathrm{RCN}$$

Scheme 15. Synthesis of nitriles using ZnO

2.13 Synthesis of α–aminophosphonates

Due to the biological activities of a-aminophosphonates, the search for new catalysts leading to an efficient and practical methodology for the synthesis of these compounds is highly desired. Kassaee et al. reported that zinc oxide nanoparticles were used as an effective catalyst in the solvent-free, three-component couplings of aldehydes, aromatic amines and dialkyl phosphites at room temperature to produce various a-amino phosphonates. Major advantages with this protocol include short reaction times, mild reaction conditions, easy workup and generality (Scheme 16) (Kassaee et al., 2009).

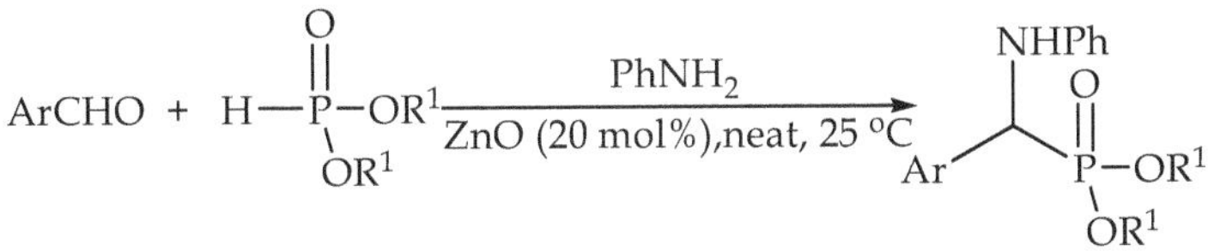

Scheme 16. ZnO nanoparticle catalyzed the synthesis of α-amino phosphonates

More recently, the author has utilized the nano ZnO catalyst for synthesis of α–aminophosphonic esters bearing a ferrocenyl moiety revealed that only few papers have been published on the synthesis of these compounds (Hosseini-Sarvari, 2011). To expand the scope of this novel transformation, the author used nano ZnO as catalyst for the synthesis of a range of new ferrocenyl aminophosphonates. A mechanistic proposal for the role of nano ZnO as the catalyst was also investigated. The activities of several other nano and bulky metal oxides and catalysts reported recently have been compared with nano ZnO. This study shows that the yield of desired product in the presence of nano ZnO is comparably higher than other catalysts used (Table 6).

Entry	Catalyst	Time (h)	Conversion (%)	Yield (%)
1	Bulky ZnO	12	90	84
2	Bulky Fe_2O_3	24	0	0
3	Bulky basic-Al_2O_3	12	49	45
4	Bulky CuO	24	6	Trace
5	Bulky MgO	24	0	0
6	Bulky CaO	24	0	0
7	Bulky TiO_2	12	56	50
8	Nano TiO_2	12	58	50
9	Nano MgO	24	0	0
10	$Mg(ClO_4)_2$	12	7	Trace
11	$H_3PW_{12}O_{40}$	24	0	0
12	No catalyst	24	4	Trace
13	Nano ZnO	2	100	95

Table 6. Comparison of catalytic activity of nano ZnO catalyst with several other catalysts for the synthesis of diethyl anilino(ferrocenyl)methyl phosphonate

2.14 Synthesis of bis(indolyl)methanes

Bis(indolyl)methanes are the most active and highly recommended cruciferous substances for promoting beneficial estrogen metabolism and inducing apoptosis in human cancer cells. The electrophilic substitution reaction of indoles with aldehydes is one of the most simple and straightforward approaches for the synthesis of bis(indolyl)methanes. Hosseini-Sarvari successfully synthesized bis(indolyl) methanes by the reaction of indole with various aldehydes in the presence of ZnO catalyst in solvent-free conditions (Scheme 17). It was observed from this study that ZnO is an efficient catalyst for the synthesis of bis(indolyl)methanes in terms of product yields, reaction temperature, and reaction times (Hosseini-Sarvari, 2008).

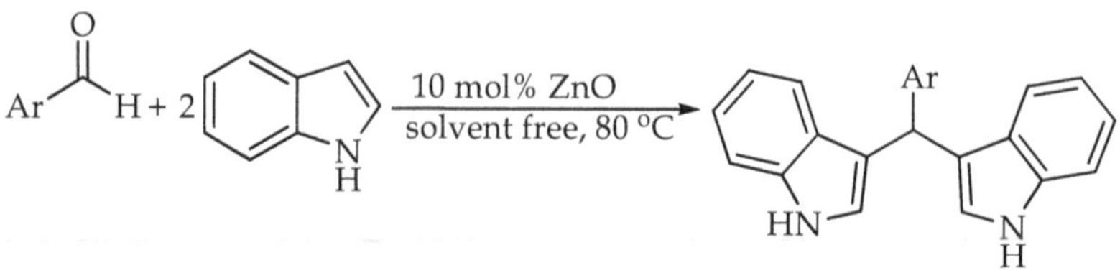

Scheme 17. Synthesis of bis(indolyl)methanes by ZnO

2.15 Synthesis of *N*-Sulfonylaldimines

N-sulfonylimines have been increasing importance because they are one of the few types of electron-deficient imines that are stable enough to be isolated but reactive enough to undergo addition reactions. ZnO was reported as a mediated for preparations of *N*-sulfonylimines under solvent-free conditions by conventional heating (Scheme 18). The advantages of this method are as follows: *i*) there is no need of toxic and waste producing Lewis acids; *ii*) work-up is simple; *iii*) the reaction procedure is not requiring specialized equipment; *iv*) zinc oxide powder can be re-used; and *v*) solvent-free condition (Hosseini-Sarvari & Sharghi, 2007).

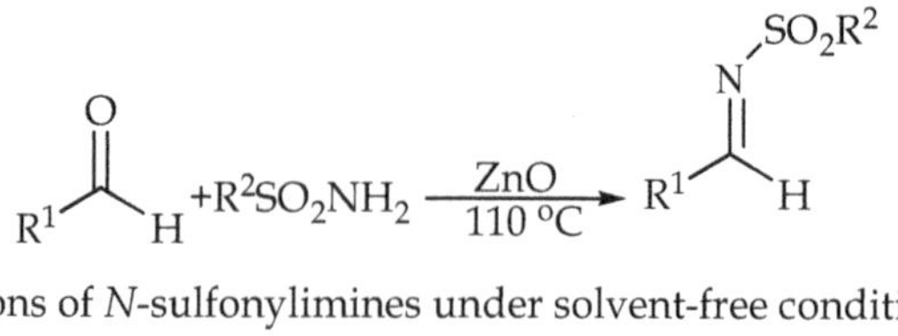

Scheme 18. Preparations of *N*-sulfonylimines under solvent-free conditions by ZnO

2.16 Synthesis of *β*–Chloro-*α*,*β*-unsaturated Ketones

A useful reaction for the synthesis of *β*–Chloro-*α*,*β*-unsaturated unsaturated ketones (as synthetic intermediates particularly for the synthesis of heterocyclic systems) involves the addition of acid chloride derivatives to terminal alkynes. However, the addition of acid chlorides to alkynes often proceeds with concomitant decarbonylation. Recently, ZnO was shown a useful catalyst for the addition of acid chlorides to terminal alkynes, afforded (Z)-adducts selectively without decarbonylation at room temperature under solvent-free conditions (Scheme 19). This protocol benefits from short reaction times, operational simplicity, neutral reaction conditions, reusability of the catalyst, avoidance of solvents, reduced environmental and economic impacts, and chemo selectivity. No toxic reagent or by product were involved and no laborious purifications were necessary (Hosseini-Sarvari & Mardaneh, 2011).

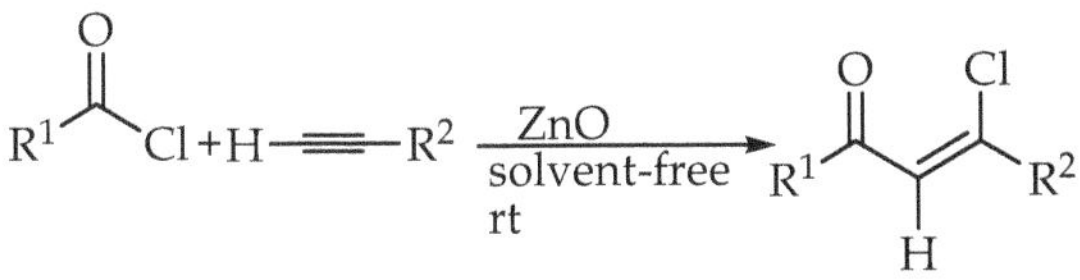

Scheme 19. Addition of acid chlorides to terminal alkynes, catalyzed by ZnO

2.17 Ring-opening of epoxides

β–Amino alcohols are synthesized by acid catalyzed ring-opening of epoxides. Hosseini-Sarvari carried out the ring-opening of cyclohexene oxide, phenoxy oxide, styrene oxide, and epichlorohydrine oxide with various aromatic amines toward the synthesis of β–amino alcohols catalyzed by ZnO, affording high yields of products under solvent-free conditions and the reaction is also regioselective (Scheme 20) (Hosseini-Sarvari, 2008).

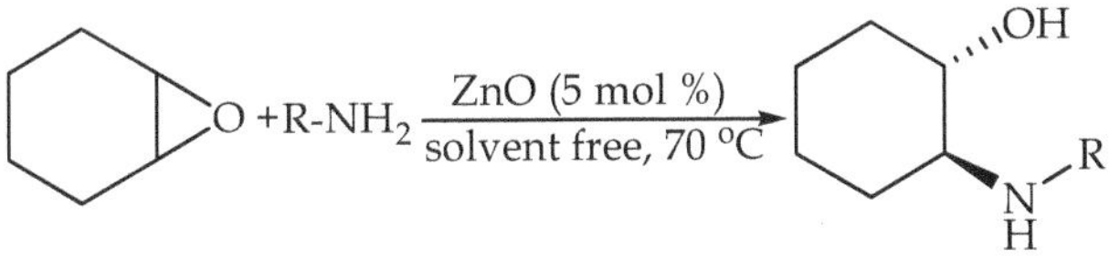

Scheme 20. Synthesis of β–amino alcohols catalyzed by ZnO

2.18 Beckmann rearrangement

The Beckmann rearrangement is a fundamental and useful reaction, long recognized as an extremely valuable and versatile method for the preparation of amides or lactams, and often employed even in industrial processes. The conventional Beckmann rearrangement usually requires the use of strong Bronsted or Lewis acids, *i.e.* concentrated sulfuric acid, phosphorus pentachloride in diethyl ether, hydrogen chloride in acetic anhydride, causing large amounts of byproducts and serious corrosion problems. The ZnO catalyzed Beckmann rearrangement of various aldehydes and ketones in solvent-free conditions (Scheme 21) in good-to-excellent yields (60-95 %) (Sharghi & Hosseini, 2002). It was found that various types of aldehydes in the presence of ZnO were condensed cleanly, rapidly and selectively with hydroxylamine hydrochloride at 80°C in 5–15 min to afford the corresponding Z-isomer of the oximes (OH *syn* to aryl) in excellent yields. Only a small amount of E-isomer, *i.e.* ca. 10–20% was obtained. (Scheme 22).

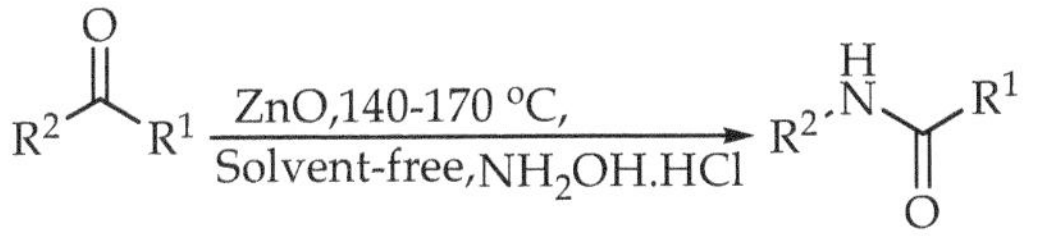

Scheme 21. Beckmann rearrangement catalyzed by ZnO

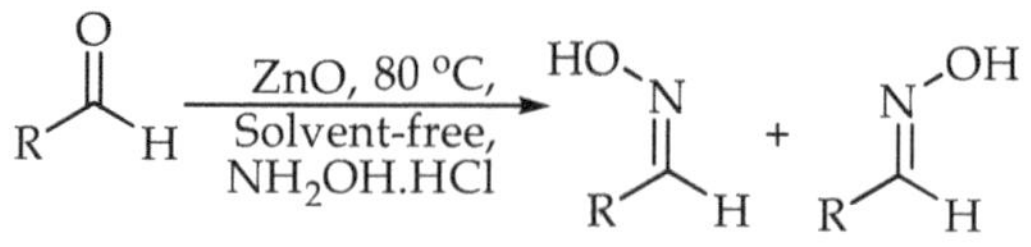

Scheme 22. Synthesis of oximes catalyzed by ZnO

2.19 Oxidation reactions

Oxygen anions on metal oxide surfaces can act as Lewis as well as Brønsted bases. As such, they may oxidize adsorbed organics. The most common examples of such reactions in the metal oxide surface science literature are nucleophilic oxidations of carbonyl compounds. Aldehydes are oxidized to the corresponding carboxylates on a number of oxide surfaces such as ZnO. Higher alcohols and aldehydes also form carboxylate intermediates on ZnO (Vohs et al., 1986, 1988, 1989). Other related species such as esters exhibit similar chemistry; oxidation of methyl formate on the ZnO (001) surface (Scheme 23).

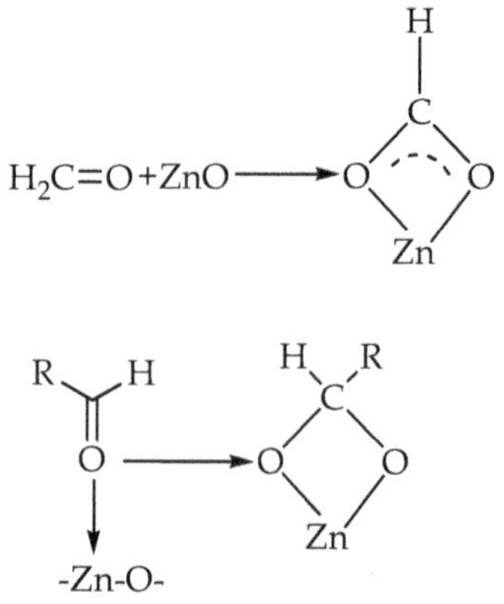

Scheme 23. Oxidation of methyl formate on the ZnO surface

Recently, a new synthetic method for the oxidation of sulfides on ZnO surface has been reported (Shiv et al., 2009) in the presence of H_2O_2 under solvent-free conditions (Scheme 24). The results reveal that PANI/ZnO composite has high activity and selectivity compared to the raw ZnO.

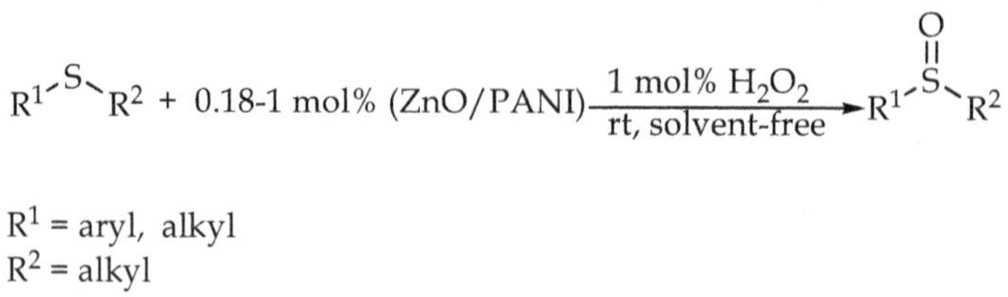

$R^1 = $ aryl, alkyl
$R^2 = $ alkyl

Scheme 24. Selective oxidation of sulfide with ZnO

3. Conclusion

This chapter review describes various organic reactions on ZnO. During the past decades, numerous organic reactions have been developed using ZnO as a non-toxic metal oxide in response to the demand for more environmentally benign organic syntheses. This development promotes the use of ZnO because of its unique properties, as described in this review. ZnO appear to be attractive for conducting organic reactions in solvent-free condition, reusability of ZnO, especially nano ZnO. When a new reaction is discovered, a devoted chemist can no longer ignore the possibility of performing the reaction using the ZnO. Thus, ZnO is a new, useful, and powerful catalyst for organic reactions.

In addition, many efforts could be found in the literature to improve the activity and stability of the ZnO catalysts, including promotion of the catalyst with transition metals like Pt, Pd, Fe, Mn, and etc. and mixed metal oxides such as TiO_2, ZrO_2, CaO, CuO, etc. which was not mentioned here. ZnO and its promoted versions are much more promising for various organic reactions of practical significance and are expected to gain great interest in the coming years.

4. Acknowledgment

The author thank the Shiraz University Research Council for their financial support.

5. References

Alinezhad, H., Salehian, F., & Biparva, P. (2012). Synthesis of benzimidazole derivatives using heterogeneous ZnO nanoparticles. *Synthetic Communication*, Vol. 42, No. 1, (September 2011), pp. 102-108, ISSN 0039-7911

Ashoka, S., Chithaiah, P., Thipperudaiah, K. V., & Chandrappa, G. T. (2010). Nanostructural zinc oxide hollow spheres: A facile synthesis and catalytic properties. *Inorganica Chimica Acta*, Vol. 363, No. 13, (October 2010), pp. 3442-3447, ISSN 0020-1693

Bahrami, K., Khodaei, M. M., & Farrokhi, A. (2009). Highly efficient solvent-free synthesis of dihydropyrimidinones catalyzed by zinc oxide. *Synthetic Communication*, Vol. 39, No. 10, (April 2009), pp. 1801-1808, ISSN 0039-7911

Bandgar, B. P., More, P. E., Kamble, V. T., & Sawant, S. S. (2009). Convenient and efficient synthesis of thiol esters using zinc oxide as a heterogeneous and eco-friendly catalyst. *Australian Journal of Chemistry*, Vol. 61, No. 12, (April 2009), pp. 1006-1010, ISSN 0004-9425

Bunn, C.W. (1935). The lattice-dimensions of zinc oxide. *Proceeding of the Physical Society*, Vol. 47, No. 5, (September 1935), pp. 835-842, ISSN 0370-1328

Catti, M., Noel, Y., & Dovesi, R. (2003). Full piezoelectric tensors of wurtzite and zinc blende ZnO and ZnS by first-principles calculations. *Journal Physics and Chemistry Solids*, Vol. 64, No. 11, (April 2003), pp. 2183-2190, ISSN 0022-3697

Dotani, M. (1997). Jpn. Kokai Tokkyo Koho, JP Patent 09048769 A 19970218

Garcı́a-Tellado, F., Loupy, A., Petit, A., & Leilani Marrero-Terrero, A. (2003). Solvent-free microwave-assisted efficient synthesis of 4,4-disubstituted 2-oxazolines.*European Journal of Organic Chemistry*, (November 2003), pp. 4387-4391, ISSN 2153-2249

Green, T. W., & Wuts, P. G. M. (1999). *Protective Groups in Organic Synthesis*, 3rd ed., Wiley–Interscience, New York, ISBN 0-471-16019-9

Hantzsch, A. (1881). Condensationprodukte aus aldehydammoniak und ketoniartigen verbindungen. *Chemische Berichte*, Vol. 14, No. 2, (July 1881), pp. 1637-1638, ISSN 0009-2940.

Heravi, M. M., Bakhtiari, K., Javadi, M. N., Bamoharram, F. F., Saeedi, M., & Oskooie, A. H. (2007). K$_7$[PW$_{11}$CoO$_{40}$]-catalyzed one-pot synthesis of polyhydroquinoline derivatives via the Hantzsch three component condensation. *Journal of the Molecular Catalysis A: Chemical*, Vol. 264, No. 1-2, (March 2007), pp. 50-52, ISSN 1381-1169

Hosseini-Sarvari, M., & Sharghi, H. (2004). Reactions on a solid surface. A simple, economical and efficient Friedel-Crafts acylation reaction over zinc oxide (ZnO) as a new catalyst. *Journal of Organic Chemistry*, Vol. 69, No. 20, (April 2004), pp. 6953-6956, ISSN 0022-3263

Hosseini-Sarvari, M., & Sharghi, H. (2005). Zinc oxide (ZnO) as a new, highly efficient, and reusable catalyst for acylation of alcohols, phenols and amines under solvent free conditions. *Tetrahedron*, Vol. 61, No. 46, (September 2005), pp. 10903-10907, ISSN 0040-4020

Hosseini-Sarvari, M. (2005). ZnO/CH$_3$COCl: A new and highly efficient catalyst for dehydration of aldoximes into nitriles under solvent-free condition. *Synthesis*, No. 5, (November 2004), pp. 787-790, ISSN 0039-7881

Hosseini-Sarvari, M., & Sharghi, H. (2006). ZnO as a new catalyst for N-formylation of amines under solvent-free conditions. *Journal of Organic Chemistry*, Vol. 71, No. 17, (April 2006), pp. 6652-6654, ISSN 0022-3263

Hosseini-Sarvari, M., & Sharghi, H. (2007). A novel method for the synthesis of N-sulfonylaldimines by ZnO as a recyclable neutral catalyst under solvent-free conditions. *Phosphorus, Sulfur, Silicon, and Related Elements*, Vol. 182, No. 9, (September 2007), pp. 2125-2130, ISSN 1042-6507

Hosseini-Sarvari, M. (2008). Synthesis of bis(indolyl)methanes using a catalytic amount of ZnO under solvent-free conditions. *Synthetic Communication*, Vol. 38, No. 6, (January 2008), pp. 832-840, ISSN 0039-7911

Hosseini-Sarvari, M., & Etmad, S. (2008). Nanosized zinc oxide as a catalyst for the rapid and green synthesis of b-phosphono Malonates. *Tetrahedron*, Vol. 64, No. 23, (March 2008), pp. 5519-5523, ISSN 0040-4020

Hosseini-Sarvari, M., Sharghi, H. & Etemad, S. (2008). Nanocrystalline ZnO for Knoevenagel condensation and reduction of the carbon, carbon double bond in conjugated alkenes. *Helvetica Chimica Acta*, Vol. 91, No. 4, (April 2008), pp. 715-724, ISSN 0018-019X

Hosseini-Sarvari, M. (2008). Synthesis of β-aminoalcohols catalyzed by ZnO. *Acta Chimica Slovenica*, Vol. 55, No.2, (June 2008), pp. 440-447, ISSN 1318-0207

Hosseini-Sarvari, M. (2011). Synthesis of quinolines using nano-flake ZnO as a new catalyst under solvent-free conditions. *Journal of Iranian Chemical Society*, Vol. 8, No. (July 2010), pp. 119-128, ISSN 1735-207X

Hosseini-Sarvari, M. (2011). An efficient and eco-friendly nanocrystalline zinc oxide catalyst for one-pot, three component synthesis of new ferrocenyl aminophosphonic esters under solvent-free condition. *Catalysis Letters*, Vol. 141, (November 2010), pp. 347-355, ISSN 1011-372X

Hosseini-Sarvari, M., & Mardaneh, Z. Selective and CO-retentive addition reactions of acid chlorides to terminal alkynes in synthesis of β-chloro-α,β-unsaturated ketones using ZnO. (2011). *Bulletin of the Chemical Society Japan*, Vol. 84, No. 7, (February 2011), pp. 778-782, ISSN 0009-2673

Ishikawa, R. (1999). JPn. Kokai Tokkyo Koho, JP Patent 11217376 A 19990810

Jackson, S. D., & Hargreaves, J. S. J. (2009). *Metal Oxide Catalysis*, 692-694, ISBN 978-3-527-31815-5

Kakiuchi, K., Hosono, E., & Fujihara, S. (2006). Enhanced photoelectrochemical performance of ZnO electrodes sensitized with N-719. *Journal of Photochemistry and Photobiology, A*, Vol. 179, Vo.1-2, (April 2006), pp.81-86, ISSN 1751-1097

Kassaee, M. Z., Movahedi, F., & Masrouri, H. (2009). ZnO nanoparticles as an efficient catalyst for the one-pot synthesis of α-amino phosphonates. *Synlett*, Vol. 8, (December 2008), pp.1326-1330, ISSN1437-2096

Kassaee, M. Z., Masrouri, H., & Movahedi, F. (2010). ZnO-nanoparticle-promoted synthesis of polyhydroquinoline derivatives via multicomponent Hantzsch reaction. *Monatshefte fur Chemie*, Vol. 141, No. 3, (January 2010), pp. 317-322, ISSN 1434-4475

Katkar, S. S., Mohite, P. H., Gadekar, L. S., Arbad, B. R., & Lande, M. K. (2010). ZnO-beta zeolite: as an effective and reusable heterogeneous catalyst for the one-pot synthesis of polyhydroquinolines. *Green Chemistry Letters and Review*, Vol. 3, No. 4, (Desember 2010), pp. 287-292, ISSN 1751-8253

Katkar, S. S., Arbad, B. R., & Lande, M. K. (2011). ZnO-beta zeolite catalyzed solvent-free synthesis of polyhydroquinoline derivatives under microwave irradiation. *Arab Journal of Science and Engineering*, Vol. 36, No. 1, (January 2011), pp. 39-46, ISSN 1319-8025

Knoevenagel, L. F. (1898). Condensation von malonsäure mit Aromatiachen aldehyden durch ammoniak und amine. *Berichte der deutschen chemischen Gesellschaft*, Vol. 31, No. 3, (October 1898), pp. 2596–2619, ISSN 0365-9631

Kong, X. Y., & Wang, Z. L. (2004). Polar-surface dominated ZnO nanobelts and the electrostatic energy induced nanohelixes, nanosprings, and nanospirals. *Applied Physics Letters*, Vol. 84, No. 6, (December 2003), pp. 975-977, ISSN: 0003-6951.

Kumar, A., & Maurya, A. R. (2007). Bakers' yeast catalyzed synthesis of polyhydroquinoline derivatives via an unsymmetrical Hantzsch reaction. *Tetrahedron Letters*, Vol. 48, No. 22, (March 2007), pp. 3887-3890, ISSN 0040-4039

Moghaddam, F. M., & Saeidian, H. (2007). Controlled microwave-assisted synthesis of ZnO nanopowder and its catalytic activity for O-acylation of alcohol and phenol. *Material Science and Engineering: B*, Vol. 139, No. 2-3, (March 2007), pp. 265–269, ISSN 0921-5107

Moghaddam, F. M., Saeidian, H. Mirjafary, Z., & Sadeghi, A. (2009). Rapid and efficient one-pot synthesis of 1,4-dihydropyridine and polyhydroquinoline derivatives through the Hantzsch four component condensation by zinc oxide. *J. Iran. Chem. Soc.*, Vol. 6, No. 2, (June 2009), pp. 317-324, ISSN 1735-207X

Morimoto, M., & Ishikawa, R. (1997). JPn. Kokai Tokkyo Koho, JP Patent 09301960 A 19971125

Oresmaa, L., Taberman, H., Haukka, M., Vainiotalo, P., & Aulaskari, P. (2007). Regiochemistry of N-substitution of some 4(5)-substituted imidazoles under solvent-free conditions. *Journal of Heterocyclic Chemistry*, Vol. 44, No. 6, (November 2007), pp. 1445-1451, ISSN 0022-152X

Reddy, M. B. M., & Pasha, M. A. (2010). Environment friendly protocol for the synthesis of nitriles from aldehydes. *Chinese Chemical Letters*, Vol. 21, No. 9, (December 2010), pp. 1025-1028, ISSN 1001-8417

Sharma, P., Gupta, A. K., Rao, V., Owens, F. J., Sharma, R., Ahuja, R., Osorio, J. M., Johansson, B., & Gehring, G. A. (2003). Ferromagnetism above room temperature in bulk and transparent thin films of Mn-doped ZnO. *Nature Material*, Vol. 2, No. 10, (September 2003), pp. 673-677, ISSN 1476-1122

Sharghi, H., & Hosseini, M. (2002). Solvent-free and one-step Beckmann rearrangement of ketones and aldehydes by zinc oxide. *Synthesis*, No. 8, (April 2002), pp. 1057-1060, ISSN 0039-7881

Shaterian, H. R., & Ghashang, M. (2008). A highly efficient method for the silylation of alcohols, phenols, and naphthols using HMDS in the presence of zinc oxide (ZnO) as economical heterogeneous catalyst. *Phosphorus Sulfur Silicon Related Elements*, Vol. 183, No. 1, (February 2008), pp. 194-204, ISSN 1042-6507

Shiv, P., Sharma, M. V. S., Suryanarayana, A. K., Nigam, A. S., Chauhan, A. S., & Tomar, L. N. S. (2009). [PANI/ZnO] composite: Catalyst for solvent-free selective oxidation of sulfides. *Catalysis Communication*, Vol. 10, No. 6, (February 2009), pp. 905-912, ISSN 1566-7367

Tammaddon, F., Amrollahi, M. A., & Sharafat, L. (2005). A green protocol for chemoselective *O*-acylation in the presence of zinc oxide as a heterogeneous, reusable and eco-friendly catalyst. *Tetrahedron Letters*, Vol. 46, No. 45, (Vovember 2005), pp. 7841-7844, ISSN 0040-4039

Tayebee, R., Cheravi, F., Mirzaee, M., & Amini, M. M. (2010). Commercial zinc oxide (Zn^{2+}) as an efficient and environmentally benign catalyst for homogeneous benzoylation of hydroxyl functional groups. *Chinese Journal of Chemisry*, Vol. 28, No. 7, (July 2010), pp. 1247–1252, ISSN 1614-7065

Thakuria, H., Borah,B. M., & Das, G. (2007). Macroporous metal oxides as an efficient heterogeneous catalyst for various organic transformations-A comparative study. *Journal of Molecular Catalysis A: Chemical*, Vol. 274, No. 1-2, (September 2007), pp. 1–10, ISSN 1381-1169

Vohs, J. M., & Barteau, M. A. (1986). Conversion of methanol, formaldehyde and formic acid on the polar faces of zinc oxide. *Surface Science*, Vol. 176, No. 1-2, (October 1986), pp. 91-114, ISSN 00396028

Vohs, J. M., & Barteau, M. A. (1988). Spectroscopic characterization of surface formates produced via reaction of HCOOH and $HCOOCH_3$ on the (0001) surface of zinc oxide. *Surface Science*, Vol. 197, No. 1-2, (January 1988), pp. 109-122, ISSN 00396028

Vohs, J. M., & Barteau, M. A. (1988). Reaction pathways and intermediates in the decomposition of acetic and propionic acids on the polar surfaces of zinc oxide. *Surface Science*, Vol. 201, No. 3, (July 1988), pp. 481-502, ISSN 00396028

Vohs, J. M., & Barteau, M. A. (1988). Alkyl elimination from aldehydes on zinc oxide: Relevance to allylic oxidation pathways. *Journal of Catalysis*, Vol. 113, No. 2, (October 1988), pp. 497-508, ISSN 0021-9517

Vohs, J. M., & Barteau, M. A. (1989). Formation of stable alkyl and carboxylate intermediates in the reactions of aldehydes on the zinc oxide (0001) surface. *Langmuir*, Vol. 5, No. 4, (July 1989), pp. 965-972, ISSN 0743-7463

Vohs, J. M., & Barteau, M. A. (1989). Activation of aromatics on the polar surfaces of zinc oxide. *Journal of Physical Chemistry: B*, Vol. 93, No. 26, (December 1989), pp. 8343-8354, ISSN 1520-6106

Wang, X. Ding, Y. Summers, C. J., & Wang, Z. L. (2004). Large-Scale Synthesis of Six-Nanometer-Wide ZnO Nanobelts. *Journal of Physical Chemistry: B*, Vol. 108, No. 26, pp. 8773-8777, ISSN 1520-6106

Wang, R., Hong, X., & Shan, Z. (2008). A novel, convenient access to acylferrocenes: acylation of ferrocene with acyl chlorides in the presence of zinc oxide. *Tetrahedron Letters*, Vol. 49, No. 4, (January 2008), pp. 636-641, ISSN 0040-4039

8

Green Synthesis and Characterizations of Silver and Gold Nanoparticles

Nora Elizondo et al.*

Facultad de Ciencias Físico-Matemáticas, N. L., CP 66451,
México

1. Introduction

Metallic nanoparticles (nps) are of great interest because of the modification of properties observed due to size effects, modifying the catalytic, electronic, and optical properties of the monometallic nps.[Bronstein et al., 2000; Chushak & Bartell, 2003; Tomas, 2003]

In the last years, biosynthesis of nps have been received considerable attention due to the growing need to develop clean, nontoxic chemicals, environmentally benign solvents and renewable materials [Gericke and Pinches, 2006; Harris and Bali, 2008]. As a result, researchers in the field of nanoparticle synthesis and assembly have turned towards the utilization of biological system such as yeast, fungi, bacteria and plant extracts for the synthesis of biocompatible metal and semiconductor nps through control nucleation and growth of inorganic nps [Kasthuri et al., 2009; Lee et al., 2011; Shankar et al., 2003].

The green method employing plant extracts have drawn attention as a simple and viable alternative to chemical procedures and physical methods, which consist of a low concentration of gold or silver precursor that is added to plant extract in solution to make up a final solution and centrifuged. The supernatant is heated at 50°C to 95°C. A change in the color of solution is observed during the heating process. Bioreduction of silver ions to yield metal nanoparticles using living plants, geranium leaf [Shankar et al., 2003], Neem leaf [Shankar et al., 2004a]. Very recently, they have demonstrated synthesis of gold nanotriangles and silver nps using *Aloevera* plant extracts [Chandran et al., 2006], *Emblica officinalis* (amla, Indian Gooseberry).[Amkamwar et al., 2005] Most of the above research on the synthesis of silver or gold nps utilizing plant extracts employed broths resulting from boiling fresh plant leaves. The green synthesis of silver nps using *Capsicum annuum* leaf extract has been reported.[Li et al., 2007] According to previous reports, the polyol components and the water-soluble heterocyclic components are mainly responsible for the

* Paulina Segovia[1,3], Víctor Coello[3], Jesús Arriaga[1], Sergio Belmares[1], Aracelia Alcorta[1],
Francisco Hernández[1], Ricardo Obregón[1], Ernesto Torres[2] and Francisco Paraguay[4]
[1]*Facultad de Ciencias Físico-Matemáticas, México*
[2]*Facultad de Medicina, México*
Universidad Autónoma de Nuevo León, San Nicolás de los Garza, N. L., México
[3]*CICESE, Monterrey, PIIT, Apodaca, N. L., México*
[4] *CIMAV, Chihuahua, Complejo Ind. Chih., Chihuahua, Chih., México*

reduction of silver ions and the stabilization of the nps, respectively[Arangasamy & Munusamy, 2008; Nagajyoti et al., 2011].

Specific synthesis of nps and nanostructured materials are attracting attention in recent research because of their valuable properties which make them useful for catalysis, [El-Sayed & Narayanan, 2004] sensor technology, [Gomez-Romero, 2001] biological labeling, [Shankar et al., 2003] optoelectronics recording media and optics.[Qiu et al. 2004] The size, shape and surface morphology play pivotal roles in controlling the physical, chemical, optical and electronic properties of these nanoscopic materials.[Gracias et al., 2002; Kamat, 2002] This is particularly important for noble metals such as Au and Ag which have strong surface plasmon resonance (SPR) oscillations. The shape-selective metal nps such as rods, tubes, wires, triangles, prisms, hexagons and cubes can be regularly synthesized by chemical, biological and physical methods. [El-Sayet, 2001; Lim et al., 2008]

Many colloidal methods of synthesis have been approached to obtain metallic nps for this purpose, such as homogeneous reduction in aqueous solutions,[Shankar et al., 2004b]or phase transfer reactions,[Liz-Marzan & Philipse, 1995] with sodium citrate, hydrazine, NaBH$_4$, and lithium triethylborohydride (LiBEt$_3$H) as reducing agents, each of them yielding products with different physicochemical and structural characteristics.[Han et al., 1998] Among these, the polyol method has been reported to produce small nps as the final product, easily changing composition and surface modifiers. This technique does not require an additional reducing agent since the solvent by itself reduces the metallic species. However, besides the stoichiometry and order of addition of reagents in the synthesis process, one of the most important parameters in the preparation is the temperature. Modifications in temperature influence the reaction by changing the stabilization of the nps formed and the surface modifiers, e.g., PVP, and the nucleation rate of the reduced metallic atoms.[Schmid, 1994]

Gold (Au) and silver (Ag) nps have a diversity of interesting properties between which they emphasize the electrical ones, optical, catalytic and the applications in biomedicine like antibacterial and antiviral, same that depend on their morphology and size.

Characterization of these systems has been a difficult process where researchers have employed indirect measurements to identify the localization of the elements within the nps. A novel approach to study this kind of particles is based on the use of a high angle annular dark field (HAADF) technique, in a transmission electron microscope (TEM), which allows the observation of the elements due to atomic number, densities, or the presence of strain fields due to differences in lattice parameters, structure, the presence of surfactants or any other surface modifier besides the size of the particle and also by near-field scanning optical microscopy (NSOM) we determine the size of the particles.[Henglein, 2000; Turkevich et al., 1951]

The nps were synthesized using polyol and green methods. We made a comparison of these methods in order to investigate the influence of reaction parameters on the resulting particle size and its distribution. In the first method we use polyol process with poly (vinylpyrrolidone) (PVP) acting as a stabilizer and ethylenglycol as a reductor.[Cao, 2004; Park et al., 2008] Such procedure yield different morphologies of metal nps (including gold and silver). [Burda et al., 2005; Gonzalez et al., 2009; Kasthuri et al., 2009; Rosi & Mirkin, 2005; Safaepour et al., 2009; Xia & Halas, 2005] The green method is an ecological synthesis technique. There, we made use of chemical compounds of plants like *Rosa Berberifolia* and *Geranium Maculatum* in order to obtain ascorbic acid and polyphenols as reductor agents. Ascorbic acid ($C_6H_8O_6$) and polyphenols like hydroxyphenol compounds are abundant components of plants. Ascorbic acid reaches a concentration of over 20 milimols in chloroplasts and occurs in all cell compartments including the cell wall. Additionally the acid has functions in photosynthesis as an enzyme cofactor (including synthesis of ethylene, gibberellins and anthocyanins) and in the control of cell growth. [Altansukha, 2010; Smirnoff & Wheeler, 2000] In nature, polyphenol is one of the most important chemicals in many reductive biological reactions widely found in plants and animals. The hydroxyphenol compounds and their derivatives could be used as versatile reducing agents for facile one-pot synthesis of gold and silver nanoparticles with diverse morphological characters. Most of the reports on the biological synthesis of metal nps utilizing plant extracts employed broths obtained from boiled fresh plant leaves. In this present study, we report on the synthesis of silver and gold nps using *Aloe Barbadensis* and *Cucurbita Digitata* extracts at 60°C. The approach is a simple, cost-effective, stable for long time, reproducible and previously unexploited method excellent for nanofabrication. These plants are predominant species in America especially in Mexico. In the case of *Aloe Barbadensis* it has been used for medical applications such as there is some preliminary evidence that *Aloe Vera* extracts may be useful in the treatment of wound and burn healing, minor skin infections, sebaceous cysts, diabetes, and elevated blood lipids in humans(from Wikipedia).[Harris & Bali 2008] These positive effects are thought to be due to the presence of compounds such as polysaccharides, mannans, anthraquinones, and lectins.[Boudreau & Beland 2006; Eshun & He 2004; King et al., 1995; Vogler & Ernst 1999]

We use for the synthesizes plants like *Aloe Barbadensis* and *cactus plants like Cucurbita Digitata* that is a reminder plant in Mexico, also with compounds that have surfactant properties like saponins. Here, we show that green method reduces the temperature requirement which is in contrast to the obtained with the polyol method. The use of these natural components allows synthesize gold and silver nps.

2. Experimental section

The polyol method was followed to obtain nps passivated with poly(vinylpyrrolidone) (PVP). Hydrogen tetrachloroaurate (HAuCl$_4$) (III) hydrate (99.99%), silver nitrate (AgNO$_3$) (99.99%), and poly (N-vinyl-2-pyrrolidone) (PVP-K30, MW = 40000) were purchased from Sigma Aldrich, and 1, 2-ethylenediol (99.95%) was purchased from Fischer Chemicals; all the materials were used without any further purification. A 0.4 g sample of Poly (N-vinyl-2-pyrrolidone) (PVP) was dissolved in 50 mL of 1,2-ethylenediol (EG) under vigorous stirring, heating in reflux, until the desired temperature was reached (working temperatures ranged from 140°C to 190°C in increments of 10°C). For the monometallic nps, a 0.1 mM aqueous solution of the metal precursor was added to the EG-PVP solution, with continuous agitation for 3 h in reflux.

The green method is an ecological synthesis technique. There, we made use of chemical compounds of plants like *Rosa Berberifolia, Geranium Maculatum, Aloe Barbadensis* and *Cucurbita Digitata* in order to obtain ascorbic acid as reductor agent from the extracts of these plants as can be seen from figure 2. Ascorbic acid ($C_6H_8O_6$) is an abundant component of plants, which reaches a concentration of over 20 milimols in chloroplasts and occurs in all cell compartments including the cell wall. We use for the synthesizes also cactus extracts with compounds that have surfactant properties like saponins.

| a) | b) | c) | d) |

Fig. 1. The synthesizes by the green chemistry method were realized using extracts of plants with scientific names of: (a) *Rosa Berberiforia*, (b) *Geranium Maculatum*, (c) *Aloe Barbadensis* and (d) *Cucúrbita Digitata*.(Images of this figure are from http://www. Google.com and http://www.Wikipedia.org)

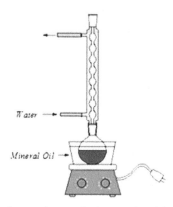

Fig. 2. Reflux system used for the synthesis of silver and gold nanoparticles by polyol and green chemistry methods.

The extracts were prepared as of 1 to 40 grams of the mentioned fresh plants. They were heated in a flask with deionized water at 100°C under stirring for 10 minutes and filtered three times. Then from 10 to 50 milliliters of the extracts of these plants respectively were dissolved in water or in ethanol under vigorous stirring, heating in reflux, until the desired temperature was reached. For the gold and silver nps, a 0.1 mM aqueous solution of the metal precursor was added to the solutions with extracts, with continuous agitation for 30 minutes to 24 h in reflux like by the polyol method as can be seen in figure 2, in a working temperatures range from 60°C to 100°C. When the precursors were added to the reaction solutions, we observed drastic changes of the color of the solutions after one minute of the reaction time from yellow to dark brown in the case of silver nps and for gold nps synthesis the color of the solutions changed from yellow-pink tones to dark brown as shown in figure 3.

The synthesis of colloidal metallic nps was carried out taking into account the optimization of the conditions of nucleation and growth. For this reason, the variation of parameters like the concentration of the metallic precursors, reductor agent, amount of stabilizer, temperature and time of synthesis were realized.

Fig. 3. Photography of monometallic colloidal dispersions of gold nanoparticles in the solutions with the extracts of *Aloe Barbadensis*, the change of color is characteristic of gold and a function of the physical properties of metallic nanoparticles obtained by green method.

For the electron microscopy analysis of the metallic nps, samples were prepared over carbon coated copper TEM grids. HAADF and HRTEM images were taken with a JEOL 2010F and a FEI TITAN microscopes in the STEM mode, with the use of a HAADF detector with collection angles from 50 mrad to 110 mrad. Also by near-field scanning optical microscopy (NSOM) we determine the size of the particles. UV-vis spectra were obtained using a 10 mm path length quartz cuvette in a Cary 5000 equipment.

3. Results and discussion

It is well known that the morphology and size distribution of metallic particles produced by the reduction of metallic salts in solution depends on various reaction conditions such as temperature, time, concentration, molar ratio of metallic salt/reducing agent, mode and order of addition of reagents, presence and type of protective agents, degree and type of agitation, and whether nucleation is homogeneous or heterogeneous [Sanguesa et al., 1992].

Following the polyol method with ethylene glycol as solvent reductor, it was possible to obtain monometallic nanoparticles with narrow size distributions in systems and different structures depending on the temperature of reaction. The monometallic synthesis of nanoparticles by itself showed distinctive morphologies of the nanoparticles depending on the temperature of reaction.

Reaction proceeds in general as an oxidation of the ethylene glycol reducing the metallic precursor to its zero-valence state. [Carotenuto et al. 2000; Sun et al., 2002]

$$OH\text{-}CH_2\text{-}CH_2\text{-}OH \rightarrow CH_3\text{-}CHO + H_2O \tag{1}$$

$$6(CH_3\text{-}CHO) + 2HAuCl_4 \rightarrow 3(CH_3\text{-}CO\text{-}CO\text{-}CH_3) + 2Au^0 + 8HCl \tag{2}$$

This reaction describes the reduction of Au^+ to Au^0.

$$OH\text{-}CH_2\text{-}CH_2\text{-}OH \rightarrow CH_3\text{-}CHO + H_2O \tag{3}$$

$$2(CH_3\text{-}CHO) + AgNO_3 \rightarrow (CH_3\text{-}CO\text{-}CO\text{-}CH_3) + Ag^0 + HNO_3 \tag{4}$$

This reaction describes the reduction of Ag+ to Ag⁰.

In the presence of a surface modifier, the reaction changes depending on the ability of the metal to coordinate with it, as in the case of PVP where the metallic precursor could coordinate with the oxygen of the pyrrolidone group, when the particles are in the nanometer size range, while when they are in the micrometer size range the coordination is mainly with the nitrogen, as reported by Bonet et al. [Bonet et al., 2000; Sun et al., 2002] as can be observed in figure 4.

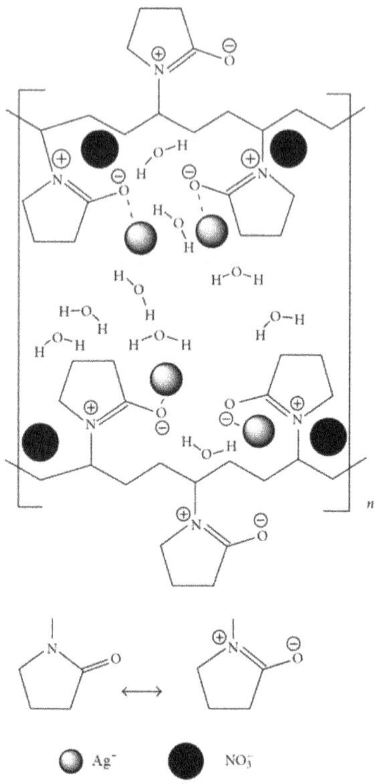

Fig. 4. A proposed mechanism of interactions between PVP and metal ions when the formed particles are in the nanometer size range.

The interaction between metal precursor and PVP has an effect on the formation of PVP-stabilized metal colloidal nanoparticles. The interaction between metal colloids and PVP is an important factor to influence the stabilities and the sizes of PVP-stabilized colloidal nanoparticles and their physicochemical properties. The reaction time for the polyol method was around 3 hours and the nps were synthesized between 100°C and 190°C. The shape and size of gold nanoparticles differs greatly from one temperature of synthesis to the next one, observing a high polydispersity in all these Au systems. The growth behavior is modified when temperature changes, allowing the presence of one-dimensional structures, spheres and angular structures. At 100°C large particles were observed in approximate sizes from 0.2 μm to 1 μm in a variety of different well-defined geometric forms such as triangles, truncated triangles, and decahedrons. Also rods with diameters between 50 and 150 nm and a few micrometers in length were observed. All of these structures had very well-defined shapes. The final product was a clear solution with large Au precipitates, some of them visible to the bare eye. At 140°C more rounded particles were observed, with shapes less defined. These particles were also smaller than in the 100°C case, approximately from 300 nm to 500 nm in sizes. At this temperature the structures observed tend to be more spherical than in the previous case. The final product at this temperature also was a clear solution with an evident precipitation of Au at the bottom of the flask. Finally, at 190°C the particles observed were smaller than in the last two cases mentioned from 200 nm to 250 nm in approximate sizes. At this temperature we can observe again particles with more geometric shapes than the ones observed at 140°C, as we can notice in Figure 5 a and b. Some rods with less than 100 nm in diameter and less than 1 μm in length were observed. The final product at this temperature had a purple color with an observable precipitation of Au at the bottom of the flask.

The reaction scheme for producing fine and monodisperse metallic nanoparticles using the polyol process involves the following successive reactions: reduction of the soluble silver nitrate and tetrachloroauric acid by ethylene glycol, nucleation of metallic silver and gold, and growth of individual nuclei in the presence of a protective agent, PVP. The fully reacted particle sizes synthesized from the polyol process depended strongly on the ramping rate of the precursor solutions to the reaction temperature; at a lower heating rate larger particles were generated, most likely due to a slower nucleation rate, while at a higher rate faster nucleation produced smaller-sized particles. At a heating rate of 2°C min^{-1}, the mean size of silver particles was 50 nm, and increasing the heating rate to 10°C min^{-1} yielded smaller and more monodisperse particles with a mean size of 25 nm as can be seen in figure 5c. The particle size of the silver decreased slightly when the reaction temperature was decreased from 150°C to 100°C.

In order to obtain monodisperse metal particles, generally, rapid nucleation in a short period of time is important; that is, almost all ionic species have to be reduced rapidly to metallic species simultaneously, followed by conversion to stable nuclei so as to be grown [Dongjo, Kim et al., 2006]. In the method of heating a precursor solution, however, both nucleation and growth can proceed gradually with increasing temperature. As such, it is difficult to synthesize particles with high monodispersity.

Therefore, the rapid injection of silver nitrate aqueous solution into ethylene glycol maintained at the reaction temperature would guarantee a short burst of nucleation after which the nuclei would continue to grow without additional nucleation, thus ensuring monodispersity.

Upon addition of the silver nitrate and tetrachloroauric acid aqueous solutions to hot ethylene glycol, the Ag$^+$ and Au$^+$ species are reduced to metallic silver and gold nanoparticles.

The concentrations of metallic silver and gold in solution increase, reaching the supersaturation conditions and finally the critical concentration to nucleate. Spontaneous nucleation then takes place very rapidly and many nuclei are formed in a short time, lowering the silver and gold concentrations below the nucleation and supersaturation levels into the saturation concentration region. After a short period of nucleation, the nuclei grow by the deposition of metallic silver and gold until the system reaches the saturation concentration. At the end of the growth period, all the metal nanoparticles have grown at almost the same rate and the systems exhibit a narrow particle size distribution.

This temperature dependence on particle size can be explained as follows. Because of the relatively high temperature used in the synthesis of silver and gold nanoparticles by polyol method, the Brownian motion and mobility of surface atoms increase. This enhances the probability of particle collision, adhesion, and subsequent coalescence. However, PVP is added to protect the particles from agglomeration. Particle coalescence is the means by which the system tries to attain thermodynamic equilibrium by reducing its total surface area.

Spherical silver nanoparticles with a controllable size and high monodispersity were synthesized by the polyol method as can be seen from figure 5c. Two different synthesis methods for producing the Ag nanoparticles were compared in terms of particle size and monodispersity. Silver nanoparticles with a size of 25 ± 4 nm were obtained at a reaction temperature of 120°C and a heating rate of 10°C min^{-1} in the precursor heating method, where the heating rate was a critical parameter affecting particle size.

In the precursor injection method, on the other hand, the injection rate and reaction temperature were important factors for producing uniform-sized Ag with a reduced size. Silver nanoparitlces with a size of 18 ± 2 nm were obtained at an injection rate of 2.0 ml s^{-1} and a reaction temperature of 100°C. The injection of the precursor solution into a hot solution is an effective means to induce rapid nucleation in a short period of time, ensuring the fabrication of silver and gold nanoparticles with a smaller size and a narrower size distribution by the polyol method.

The effect of temperature in the polyol method is crucial because at lower temperatures the oxidation potential of ethylene glycol is bigger than at higher temperatures. This means that at lower temperatures the oxidation of ethylene glycol is less favored, which traduces in fewer electrons available in the reaction environment to reduce the metals. As the temperature keeps increasing, the oxidation potential of the ethylene glycol decreases, indicating that this electrochemical reaction is favored at higher temperatures. This translates as an increment in the electrons concentration in the reaction environment. [Bonet et al., 1999]

In contrast, the reduction potential of the metals is not affected by the temperature, according to Bonet et al. It is insensitive to the reaction temperature, but there still is an effect on the reduction of the precursors related to the temperature dependence of the diffusion of metal species. Another effect of temperature on the reduction of the metal precursors is that the energy barrier that opposes to the reduction of the precursor is equal to the difference between the oxidation potential of the ethylene glycol and the reduction potential of the metal species. [Bonet et al., 1999]

Once the oxidation potential of the ethylene glycol is lowered down to the same value of the reduction potential of the metal precursor, the reduction of the metal precursor will occur spontaneously and followed by the nucleation of metal nanoparticles. [Bonet et al., 1999] From this analysis one can conclude that at higher temperatures the reduction of the metal species will be favored and also the oxidation of the ethylene glycol. This will decrease the time needed to reduce the metal precursor and the nucleation time needed to the formation of metal particles.

From this study it was found that by the polyol method the temperature plays a decisive role in the synthesis of gold and silver nanoparticles protected with PVP. It does not only affect the rates of reduction and nucleation of the metals, but it also affects the coordination between the metals and the polymeric protective agent, the distribution of elements in the nanoparticles, and the final particle size.

In the green method the reaction time is reduced from 3 hours to 30 minutes until 1 hour at 60°C, but we carried out the reaction during 24 hours in order to observe the growth of the particles.

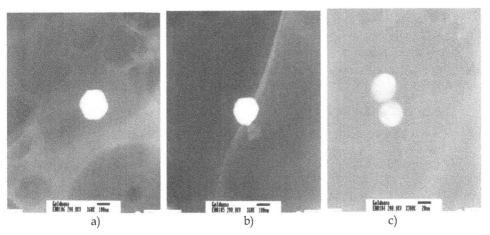

a)   b)   c)

Fig. 5. TEM images of gold nanoparticles synthesized at 190 °C (a) 200 nm in size and (b) 250 nm in size and silver nanoparticles synthesized at 120°C (c) by the polyol method.

The TEM characterization reveal the formation of nps of these metals, independent of the employed method, with a size distribution between 20 and 120 nm for gold (see figure 5) and between 10 and 27 nm for the silver.

The NSOM showed that the size of gold nps synthesized   was of 25 nm with very narrow distribution.

Plants contain a complex network of antioxidant metabolites and enzymes that work together to prevent oxidative damage to cellular components. Isolated quercetin [Wu, 2008] and polysaccharides [Ahmad et al, 2009; Collera et al., 2005; Vedpriya, 2010; Jagadeesh et al., 2004] have been used for the synthesis of silver and gold nanoparticles. Plants Extracts like *Aloe Barbadensis* is reported to contain chemically different groups of compounds: polyphenols, flavonoids, sterols, triterpenes, triterpenoid saponins, beta-phenylethylamines, tetrahydroisoquinolines, reducing sugars like glucose and fructose, and proteins, in all extracts.

The plant extract is reported to have activities of scavenging superoxide anion radicals and 1, 1-diphenyl-2-picrylhydrazyl radicals (DPPH). It could be that these water-soluble scavenging superoxide anion radicals and 1, 1-diphenyl-2-picrylhydrazyl (DPPH) radicals present in the plant extract be responsible for the reduction of silver and synthesis of nanoparticles through biogenic routes. The exact mechanism of the formation of these nanoparticles in these biological media is unknown. Presumably, biosynthetic products or reduced cofactors play an important role in the reduction of respective salts to nanoparticles. However, it seems probable that some glucose and ascorbate reduce $AgNO_3$ and $HAuCl_4$ to form nanoparticles. [Ahmad et al. 2011; Hu et al., 2003]

The probability of reduction of $AgNO_3$ to silver may be illustrated due to the mechanism known as glycolysis. Plants fix CO_2 in presence of sunlight. Carbohydrates are the first cellular constituent formed by the photosynthesizing organism on absorption of light. This carbohydrate is utilized by the cell as glucose by Glycolysis. This is the metabolic pathway that converts glucose $C_6H_{12}O_6$ into pyruvate and hydrogen ion:

$$CH_3COCOO^- + H^+ \qquad (5)$$

The free energy released in this process is used to form the high-energy compounds, ATP adenosine triphosphate and NADH (reduced nicotinamide adenine dinuleotide). Glycolysis can be represented by the following simple equation:

$$Glucose + 2ADP + 2Pi + 2NAD^+ = 2\ Pyruvate + 2ATP + 2\ NADH + 2H^+ \qquad (6)$$

Glycolysis is a definite sequence of ten reactions involving ten intermediate compounds [Ahmad et al. 2011]. Large amount of H^+ ions are produced along with ATP. Nicotinamide adenine dinucleotide, abbreviated NAD^+, is a coenzyme found in all living cells. NAD is a strong reducing agent. NAD^+ is involved in redox reactions, carrying electrons from one reaction to another. The coenzyme is therefore found in two forms in cells. NAD^+ is an oxidizing agent—it accepts electrons from other molecules and becomes reduced. This reaction forms NADH, which can donate electrons. These electron transfer reactions are the main function of NAD:

$$AgNO_3 \rightarrow Ag^+ + NO_3^-\ \ or\ \ 2HAuCl_4 \rightarrow 2Au^+ +\ \ 4HCl$$

$$NAD^+ + e^- \rightarrow NAD,$$

$$NAD + H^+ \rightarrow NADH + e^-, \qquad (7)$$

$$e^- + Ag^+ \rightarrow Ag^0\ \ or\ \ e^- + Au^+ \rightarrow Au^0$$

NAD^+ keeps on getting reoxidized and gets constantly regenerated due to redox reactions. This might have led to transformations of Ag or Au ions to Ag^0 or Au^0. Another mechanism for the reduction of Ag or Au ions to silver or gold could be due to the presence of water-soluble antioxidative substances like ascorbate. This acid is present at high levels in all parts of plants. Ascorbic acid is a reducing agent and can reduce, and thereby neutralize, reactive oxygen species leading to the formation of ascorbate radical and an electron.

This free electron reduces the Ag^+ or Au^+ ions to Ag^0 or Au^0 as can be seen in scheme 1.

In accordance with the studies of UV visible spectroscopy, whose plasmons are in figure 6 for Au and Ag synthesized nps the results shown an absorption energy in 547 nm and 415 nm respectively.

The use of these natural components allows synthesize metallic nps. In the green method, gold and silver nps were prepared by the same reduction of $HAuCl_4$ and $AgNO_3$

respectively using extracts of plants, ascorbic acid and polyphenols as reducing agents obtained from *Geranium Maculatum* leaves and *Rosa Berberiforia* petals and like natural surfactants saponins and simultaneous reducing agents in some cases were used *Aloe Barbadensis* and cactus extracts from *Cucúrbita Digitata*.

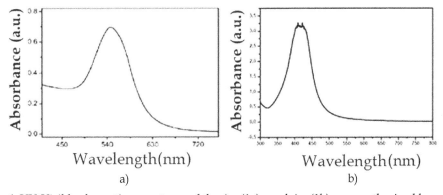

Scheme 1. Ascorbic acid reduction mechanism of gold and silver ions to obtain Ag⁰ and Au⁰ nps.

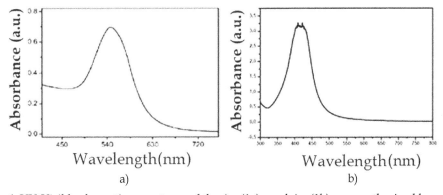

Fig. 6. UV-Visible absorption spectrum of the Au (1a) and Ag (1b) nps synthesized by polyol and green chemistry respectively.

It is important to know the exact nature of the silver and gold nanoparticles formed, and this can be deduced from the XRD Spectrum of the Sample. XRD patterns of derived Ag nps from Figure 7(a) show four intense peaks in the whole spectrum of $2\theta°$ values ranging from 20° to 90°. XRD spectra of pure crystalline silver structures have been published by the Joint Committee on Powder Diffraction Standards (file no. 04-0787). A comparison of our XRD spectrum with the Standard confirmed that the silver nanoparticles formed in our experiments were in the form of face centered cubic nanocrystals, as evidenced by the peaks at 2θ values of 38.52°, 44.49°, 64.70°, and 77.63°, corresponding to [111], [200], [220], and [311] planes for silver, respectively. In the case of gold nanoparticles in the whole spectrum of 2θ° values ranging from 35° to 80°, four new reflection signals appear at ca. 38.10°, 44.40°, 64.87°, and 77.84° in the XRD pattern of the Au, corresponding to the [111], [200], [220] and [311] planes of the Au, respectively as can be seen in Figure 7 (b), indicating that crystal structure of the gold nanoparticles was face centered cubic(JCPDS 4-0783)in this case also.

Scherrer's equation for broadening resulting from a small crystalline size, the mean, effective, or apparent dimension of the crystal composing the powder is:

$$Phkl = k\lambda/\beta 1/2 \cos\theta \qquad (8)$$

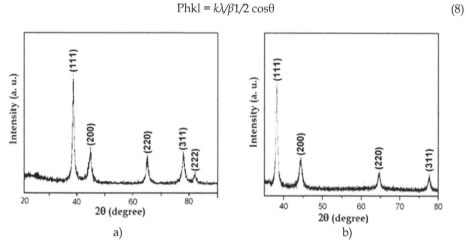

a)   b)

Fig. 7. X-ray diffractograms of silver (a) and gold (b) nanoparticles synthesized as of extracts of *Aloe Barbadensis* at 1 hour and 60°C.

where θ is the Bragg angle, λ is the wavelength of the X ray used, β is the breadth of the pure diffraction profile in radians on 2θ scale, and k is a constant approximately equal to unity and related both to the crystalline shape and to the way in which θ is defined. The best possible value of k has been estimated as 0.89. The Full Width at Half Maximum (FWHM) values measured for [111], [200], [220], and [311] planes of reflection were used with the Debye-Scherrer equation (8) to calculate the size of the nanoparticles. [Ahmad, N. et al. 2011] Moreover, the average size of the gold nanoparticles was also determined from the width of the reflection according to the Scherrer formula. The value of D calculated from the (111) reflection were k is 0.90 of the cubic phase of Au was ca. 25 nm, which is basically in agreement with the results of transmission electron microscopy (TEM) experiments for *Aloe Barbadensis* at 1 hour and 60°C.

Further analysis of the silver and gold nanoparticles by energy dispersive spectroscopy confirmed the presence of the signals characteristic of silver and gold respectively. Figure 8 shows the Energy-Dispersive Absorption Spectroscopy photographs of derived Ag nps and Au nps. All the peaks of Ag and Au respectively are observed and are assigned. Peaks for Cu and C are from the grid used, and the peaks for S, P, and Si (in the case of Au) correspond to the protein capping over the Ag nps and Au nps.

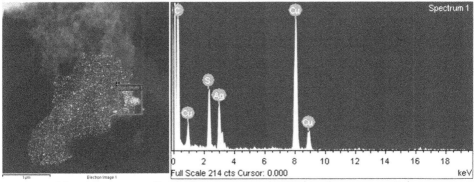

a) Image corresponding to Spectrum 1 and EDX for silver nanoparticles.

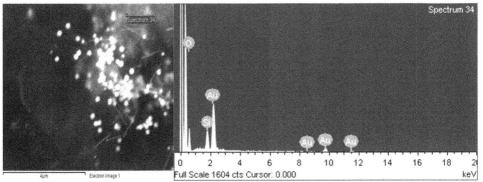

b) Image corresponding to Spectrum 34 and EDX bright field for gold nanoparticles.

Fig. 8. a) Image corresponding to select area 1 and Energy-Dispersive Absorption Spectroscopy photograph for silver nanoparticles, and b) Image corresponding to select area 34 and Energy-Dispersive Absorption Spectroscopy photograph for gold nanoparticles.

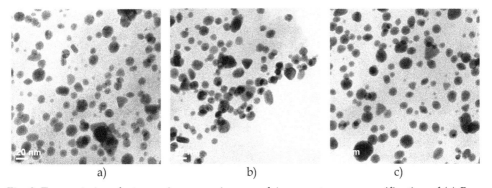

a)   b)   c)

Fig. 9. Transmission electron microscopy images of Au nps at same magnification of (a) *Rosa Berberiforia*, (b) *Geranium Maculatum*, and (c) *Cucúrbita Digitata* using the same low concentration of plants extracts approximately (0.002M) with HAuCl$_4$ (1 X 10^{-3} M) at 1 hours and 60°C.

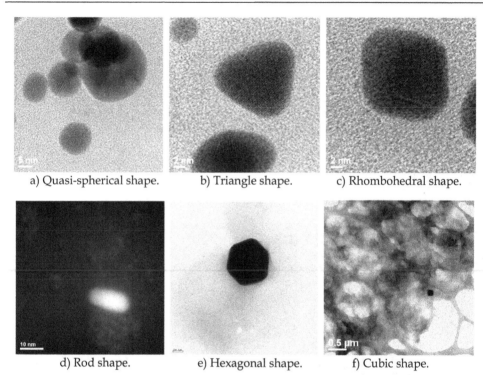

a) Quasi-spherical shape.   b) Triangle shape.   c) Rhombohedral shape.

d) Rod shape.   e) Hexagonal shape.   f) Cubic shape.

Fig. 10. Images of gold nanoparticles observed with different shapes synthesized as of *Aloe Barbadensis* extracts at different conditions varying the concentration of the extract from 0.0015 to 0.004 M, using: high resolution TEM: a) Quasi-spherical, b) Triangle, c) Rhombohedral shapes; HAADark Field TEM image: d) Rod shape; and Bright field TEM images: e) hexagonal shape and f) cubic shape gold nanoparticles.

In the figures 9 and 10, it is possible to identify large population of polydispersed gold nps synthesized as of *Rosa Berberifonia* petals, *Geranium Maculatum* leads, *Cucúrbita Digitata* cactus at same reaction conditions, and *Aloe Barbadensis* varying the concentration of plant extracts from 0.0015 to 0.004M, the consisted of spherical-, quasi-spherical-, ellipsoidal-, triangular-, hexagonal-, rombohedral-, trapezhoidal- and rod-shaped with irregular contours.

The morphology of the Ag nps was predominantly spherical and quasi-spherical as shown in figure 11, and they appear to be monodisperse for *Aloe Barbadensis* with $AgNO_3$ (1 X 10^{-3} M) at 1 hours and 60°C at different concentration of plant extract. Some of the nps were found to be oval and/or elliptical at high concentration of plant extract. Such variation in shape and size of nanoparticles synthesized by biological systems is common.

The figure 12. shows high resolution transmission electron micrographs of gold nps, synthesized with extracts of *Aloe Barbadensis* at 1 hour and 60°C using a concentration of plant extract of 0.0025 M approximately with an average size distribution of 25nm (figure 12a), the figure 12b shows a gold nanoparticle of 150 nm in size, synthesized with extracts of *Cucúrbita Digitata* using a concentration of plant extract of 0.0018M approximately and figure 12c a gold nanoparticle of 4nm in size, synthesized with extracts of *Cucúrbita Digitata* at high concentration of plant extract of 0.003M. The gold nps synthesized with extracts of *Aloe Barbadensis* were used to prepare nanoarrays for the study of optical plasmonic phenomena in another work. [Coello et al., 2010]

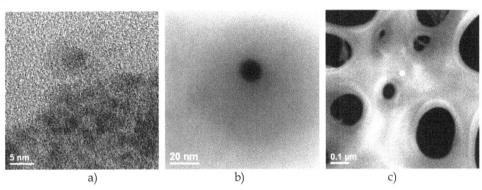

a)   b)   c)

Fig. 11. Transmission electron microscopy images of Ag nps at different magnifications of (a) High Resolution -TEM image at a concentration of plant extract of 0.004 M, (b) High-TEM image at a concentration of plant extract of 0.002 M, (c) HAADF image at a concentration of plant extract of 0.0015 M approximately of *Aloe Barbadensis* with AgNO₃ (1 X 10⁻³ M) at 1 hours and 60°C.

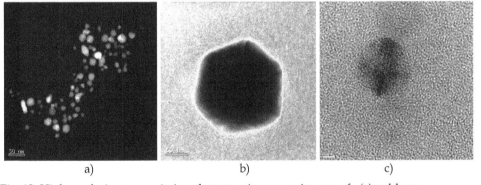

a)   b)   c)

Fig. 12. High resolution transmission electron microscopy images of   (a) gold nps synthesized with extracts of *Aloe Barbadensis* at 1 hour and 60°C using a concentration of plant extract of 0.0025 M approximately with a size distribution of 25 nm approximately, (b) a gold nanoparticle synthesized with extracts of *Cucúrbita Digitata* using a concentration of plant extract of 0.0018M approximately with a size of 150 nm and (c) a gold nanoparticle synthesized with extracts of *Cucúrbita Digitata* with a size of 4 nm at 60°C using a high concentration of plant extract of 0.003M.

The anisotropic gold and spherical–quasi-spherical silver nps were synthesized by reducing aqueous chloroauric acid (HAuCl₄) and silver nitrate (AgNO₃) solution with the extract of *Aloe Barbadensis* at 60°C temperature. The size and shape of the nps can be controlled by varying the concentration of plants extracts like *Aloe Barbadensis*.

The case of low concentration of extract with HAuCl₄ offers the aid of electron-donating group containing extract leads to formation of hexagonal-or triangular-shaped gold nps. Transmission electron microscopy (TEM) analysis revealed that the shape changes on the gold nps from hexagonal to spherical particles with increasing initial concentration of *Aloe Barbadensis*.

The electron-donating methoxy (–OCH₃) groups containing *Aloe Barbadensis* can provide a suitable environment for the formation of nps. A bioreductive approach of anisotropic gold and silver nps utilizing the *Aloe Barbadensis* has been demonstrated which provides a simple and efficient way for the synthesis of nanomaterials with tunable optical properties directed by particle shape.

The presence of small amount of *Aloe Barbadensis* leads to slow reduction of HAuCl₄ ions which facilitated the formation of triangular- or hexagonal-shaped nps. Whereas   greater amount of *Aloe Barbadensis* leads to higher population of spherical nps and was confirmed from the UV–visible and TEM analysis. The electron-donating nature of –OCH₃ group of the *Aloe Barbadensis* plays a leading role for the formation and stabilization of nps, respectively results in accordance with Kasthuri et al.[Kasthuri, et al., 2009] as shown in scheme 2.

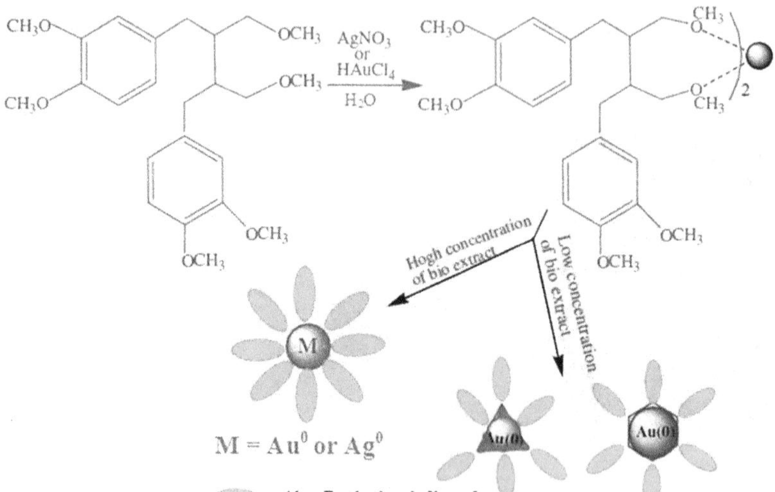

Scheme 2. The presence of small amount of *Aloe Barbadensis* leads to slow reduction of HAuCl₄ ions which facilitated the formation of triangular- or hexagonal-shaped nps. Whereas   greater amount of *Aloe Barbadensis* leads to higher population of spherical nps and was confirmed from TEM analysis.

4. Conclusion

One-step green synthesis of gold (Au) and silver (Ag) nanostructures is described using naturally occurring biodegradable plant-based surfactants, without any special reducing agent/capping agents. This green method uses water as a benign solvent and surfactant/plant extract as a reducing agent. Depending upon the Au and Ag concentration used for the preparation and the temperature, Au and Ag crystallizes in different shapes and sizes to form spherical in the case of Ag, prisms, and hexagonal structures in the case of Au. Sizes vary from the nanometer to micrometer scale level depending on the plant extract used for preparation. Synthesized Au and Ag nanostructures were characterized using scanning electron microscopy, transmission electron microscopy, X-ray diffraction, and UV spectroscopy.

In this original work, we show that green method reduces the temperature requirement, which is in contrast to the obtained with the polyol method. In the green method the size and shape of the nps can be controlled by varying the concentration of plant extracts and the reaction time. The use of these natural components allows synthesize metallic nps with very narrow distribution.

5. Acknowledgment

Authors would like to acknowledge to Facultad de Ciencias Físico Matemáticas and Microscopy Laboratory of CIIDIT de la Universidad Autónoma de Nuevo León, to Nanotechnology Laboratory of CIMAV Chihuahua, México.

6. References

Ahmad, N.; Alam, M. K.; Singh, V. N. and Sharma, S. *Journal of Bionanoscience*, 2009, 3, 2, 97–104.

Ahmad, N.; Sharma, S.; Singh, V. N.; Shamsi, S. F.; Fatma, A.; Mehta, B. R.; *Biotechnology Research International*, 2011, Article ID 454090,1- 8.

Alvarez, M. M.; Khoury, J. T.; Schaaff, G.; Shafigullin, M. N.; Vezmar, I.; Whetten, R. L. *J. Phys. Chem. B* 1997, *101*, 3706.

Altansukha, B.; Burmaa, G.; Zhianshi, J.; Van, Dan; Antsiferova, S. A. *Theoretical Foundations of Chemical Engineering*, 2010, *44, 4*, 511.

Amkamwar, B.; Damle, C.; Ahmad, A.; Sastry.; M. *J. Nanosci. Nanotechnol.* 2005, 5,1665.

Arangasamy, L.; Munusamy, V.; *Afr. J. Biotech.* 2008, 7, 3162.

Bonet, F.; Tekaia-Elhsissen, K.; Sarathy, K. V. *Bull. Mater. Sci.* 2000, *23*, 165.

Bonet, F.; Guery, C.; Guyomard, D.; Herrera-Urbina, R.; Tekaia-Elhsissen, K.; Tarascon, J. M. *Intl. J. of Inorg. Mater.* 1999, *1*, 47.

Boudreau, M.D.; Beland, F.A. *Journal of environmental science and health. Part C, Environmental carcinogenesis & ecotoxicology reviews*, 2006, *24, 1*, 103–54.

Bronstein, L. M.; Chernyshov, D. M.; Volkov, I. O.; Ezernitskaya, M. G.; Valetsky, P. M.; Matveeva, V. G.; Sulman, E. M. *J. Catal.* 2000, *196*, 302.

Burda, C.; Chen, X.; Narayanan, R.; El-Sayed, M. A. *Chem. Rev.,* 2005, *105,* 1025.

Cao, G. Nanostructures and Nanomaterials, synthesis, properties and applications, Imperial College Press, 2004.

Carotenuto, G.; Pepe, G. P.; Nicolais, L. *Eur. Phys. J. B* 2000, *16*, 11.

Chandran, S.P.; Chaudhary, M.; Pasricha, R.; Ahmad, A.; Sastry, M. *Biotechnol. Prog.* 2006, 22, 577.

Chushak; Y. G.; Bartell, L. S. *J. Phys. Chem B* 2003, *107*, 3747.

Coello V., Cortes R., Segovia P., Garcia C. and Elizondo N., *In Plasmons: Theory and Applications*, Editor: Kristina N. Helsey, Chapter 9, 2010, Nova Science Publishers.

Collera-Zuniga O., Garcia Jimenez F., and Melendez Gordillo R., *Food Chemistry*, 2005, 90, 1-2, 109–114.

Dongjo, Kim; Sunho, J.; Jooho, M., Nanotechnology, 2006, *17*, 4019–4024.

EI-Sayed, M.A.; *Acc Chem Res*, 2001, 34, 257–264.

Eshun, K.; He, Q. *Critical reviews in food science and nutrition*, 2004, 44, 2, 91–6.

Gericke, M.; Pinches, A. *Hydrometallurgy*, 2006, *83*, 132–140.

Gomez-Romero, P., *Adv. Mater*, 2001, *13*, 163–174.

González, C. M.; Liu, Y; Scaiano, J. C. *J. Phys. Chem. C*, 2009, *113*, 11861.

Gracias, D.H.; Tien, J.; Breen, T.; Hsu, C.; Whitesides, G. M.; *Science*, 2002, *289*,1170–1172.

Han, S. W.; Kim, Y.; Kim, K. *J. Colloid Interface Sci.* 1998, *208*, 272.

Harris, A.T.; Bali, R.; *J. Nanopart. Res.* 2008, *10*, 691–695.

Henglein, A. *J. Phys. Chem. B* 2000, *104*, 2201.

Hu, Y.; Xu, J.;Hu, Q.; *J. Agric. Food Chem.*2003,*51*,7788-7791.

Jagadeesh, B. H.; Prabha, T. N.; Srinivasan, K. *Indian Journal of Plant Physiology*, 2004, 9, 2,164–168.

Kamat, P.V.; *J. Phys. Chem. B* 2002, *106*, 7729–7744.

Kasthuri, J.; Kathiravan, K.; Rajendiran, N. *J Nanopart Res*, 2009, 11,1075–1085.

King, G.K.; Yates, K.M.; Greenlee, P.G. *et al. Journal of the American Animal Hospital Association*, 1995, *31*, 5, 439–47.

Lee, Y.; Gwan-Park, T. *Langmuir*, 2011, *27*, 2965–2971.

Li, S.; Shen, Y.; Xie, A.; Yu, X.; Qiu, L.; Zhang, L.; Zhang, O. *Green. Chem.* 2007, 9, 852.

Lim, J.K.; Kim, Y.; Lee, S.Y.; Joo, S.W.; 2008, *Spectrochim Acta A*, *69*, 286–289.

Liz-Marzan, L. M.; Philipse, A. P. *J. Phys. Chem.* 1995, *99*, 15120.

Nagajyoti, P.C.; Prasad,T.N.V.K.V; Sreekan, T.V.M; Lee, K. D. *Digest Journal of Nanomaterials and Biostructure.* 2011, 6, 1, 121–133.

Narayanan, R.; EI-Sayed, M.A. *Nano Lett.* 2004, *4*, 1343–1348.

Park, H. K.; Lim, Y. T.; Kim, J. K.; Park, H. G.; Chung, B. H. *Ultramicroscopy*, 2008, *108,10,* 1115.

Qiu, H.; Rieger, B; Gilbert, R.; Jerome, C.; *Chem. Mater.* 2004, *16*, 850–856.

Rosi, N. L.; Mirkin, C. A. *Chem. Rev.*, 2005, *105*, 1547.

Safaepour , M.; Reza Shahverdi, A.; Reza Shahverdi, H.; Reza Khorramizadeh, M.; Reza Gohari, A. *Avicenna J. Med. Biotech.*, 2009, *1*, 2, 111. B.

Sanguesa, C. D.; Urbina, R. H.; Figlarz, M. *J. Solid State* Chem. 1992, *100*, 272.

Schmid, G. *Clusters and Colloids, From theory to Applications*; VCH Publishers: Weinheim, Germany, 1994.

Shankar, S. S.; Ahmad, A.; Pasricha, R.; Sastry, M.; *J. Mater. Chem.* 2003, *13*, 1822–1826.

Shankar, S.S.; Rai, Ahmad, A. A.; Sastry, M. *J. Colloid. Interface. Sci.* 2004a, 275, 496.

Shankar, S. S.; Rai, A.; Ankamwar, B.; Singh, A.; Ahmad, A.; Sastry, M.; 2004b, *Nat. Mater. 3*, 482–488.

Smirnoff, N.; Wheeler G. L.; *Critical Reviews in Biochemistry and Molecular Biology*, 2000, 35, 4, 291.

Sun, Y.; Yin, Y.; Mayers, B. T.; Herricks, T.; Xia, Y. *Chem. Mater.* 2002, *14*, 4736.

Thomas, J. M.; Raja, R.; Johnson, B. F. G.; Hermans, S.; Jones, M. D.; Khimyak, T. *Ind. Eng. Chem. Res.* 2003, *42*, 1563.

Turkevich, J.; Stevenson, P.; Hillier, J. *Discuss. Faraday Soc.* 1951, *11*, 55.

Vedpriya, A.; *Digest Journal of Nanomaterials and Biostructures*, 2010, 5, 1, 9–21.

Vogler, B.K.; Ernst, E. *The journal of the Royal College of General Practitioners*, 1999, *49*, 447, 823–8.

Wu, T.H.; Yen, F.L.; Lin, L.T.; Tsai, T.R.; Lin, C.C.; Cham, T.M., *International Journal of Pharmaceutics*, 2008, 346, 1-2, 160–168.

Xia, Y.; Halas, N. J. *Mater. Res. Soc. Bull.*, 2005, *30*, 338.

Xia, Y. ; Yang, P.; Sun, Y.; Wu,Y.; Mayers, B.; Gates, B.; Yin, Y.; Kim, F.; Yan, H. *Adv. Mater.*, 2003,15, 353.

Permissions

The contributors of this book come from diverse backgrounds, making this book a truly international effort. This book will bring forth new frontiers with its revolutionizing research information and detailed analysis of the nascent developments around the world.

We would like to thank Professor M. Kidwai, Ph.D, FEnA and Dr. Neeraj Kumar Mishra, M.Sc., Ph.D, for lending their expertise to make the book truly unique. They have played a crucial role in the development of this book. Without their invaluable contribution this book wouldn't have been possible. They have made vital efforts to compile up to date information on the varied aspects of this subject to make this book a valuable addition to the collection of many professionals and students.

This book was conceptualized with the vision of imparting up-to-date information and advanced data in this field. To ensure the same, a matchless editorial board was set up. Every individual on the board went through rigorous rounds of assessment to prove their worth. After which they invested a large part of their time researching and compiling the most relevant data for our readers. Conferences and sessions were held from time to time between the editorial board and the contributing authors to present the data in the most comprehensible form. The editorial team has worked tirelessly to provide valuable and valid information to help people across the globe.

Every chapter published in this book has been scrutinized by our experts. Their significance has been extensively debated. The topics covered herein carry significant findings which will fuel the growth of the discipline. They may even be implemented as practical applications or may be referred to as a beginning point for another development. Chapters in this book were first published by InTech; hereby published with permission under the Creative Commons Attribution License or equivalent.

The editorial board has been involved in producing this book since its inception. They have spent rigorous hours researching and exploring the diverse topics which have resulted in the successful publishing of this book. They have passed on their knowledge of decades through this book. To expedite this challenging task, the publisher supported the team at every step. A small team of assistant editors was also appointed to further simplify the editing procedure and attain best results for the readers.

Our editorial team has been hand-picked from every corner of the world. Their multi-ethnicity adds dynamic inputs to the discussions which result in innovative outcomes. These outcomes are then further discussed with the researchers and contributors who give their valuable feedback and opinion regarding the same. The feedback is then collaborated with the researches and they are edited in a comprehensive manner to aid the understanding of the subject.

Apart from the editorial board, the designing team has also invested a significant amount of their time in understanding the subject and creating the most relevant covers. They scrutinized every image to scout for the most suitable representation of the subject and create an appropriate cover for the book.

The publishing team has been involved in this book since its early stages. They were actively engaged in every process, be it collecting the data, connecting with the contributors or procuring relevant information. The team has been an ardent support to the editorial, designing and production team. Their endless efforts to recruit the best for this project, has resulted in the accomplishment of this book. They are a veteran in the field of academics and their pool of knowledge is as vast as their experience in printing. Their expertise and guidance has proved useful at every step. Their uncompromising quality standards have made this book an exceptional effort. Their encouragement from time to time has been an inspiration for everyone.

The publisher and the editorial board hope that this book will prove to be a valuable piece of knowledge for researchers, students, practitioners and scholars across the globe.

List of Contributors

Ricardo Menegatti
Universidade Federal de Goiás, Brazil

Develter Dirk
Ecover Coordination Center NV, Belgium

Malaise Peter
Meta Fellowship npo, Belgium

Hong Liu and Chuan Wang
Chongqing Institute of Green and Intelligent Technology, Chinese Academy of Sciences, Chongqing, China
School of Chemistry and Chemical engineering, Sun Yat-sen University, Guangzhou, China

Yuan Liu
Chongqing Institute of Green and Intelligent Technology, Chinese Academy of Sciences, Chongqing, China

Iliana Medina-Ramirez
Department of Chemistry, Universidad Autonoma de Aguascalientes, Aguascalientes, Mexico

Maribel Gonzalez-Garcia
Department of Chemistry, Texas A&M University-Kingsville, Kingsville, TX, USA

Srinath Palakurthi
Department of Pharmaceutical Sciences, Texas A&M Health Science Center, Kingssville, TX, USA

Jingbo Liu
Nanotech and Cleantech Group, Texas A&M University-Kingsville, TX, USA
Department of Chemistry, Texas A&M University, College Station, TX, USA

Lucas Pizzuti and Márcia S.F. Franco
Universidade Federal da Grande Dourados, Mato Grosso do Sul, Brazil

Alex F.C. Flores
Universidade Federal de Santa Maria, Rio Grande do Sul, Brazil

Frank H. Quina
Instituto de Química, Universidade de São Paulo, São Paulo, Brazil

Claudio M.P. Pereira
Universidade Federal de Pelotas, Rio Grande do Sul, Brazil

Arkadiy Zhukov and Salavat Zaripov
R&D Center, GC «Mirrico», Russian Federation

Mona Hosseini-Sarvari
Department of Chemistry, Faculty of Science, Shiraz University, Shiraz, I. R. Iran

Nora Elizondo
Facultad de Ciencias Físico-Matemáticas, N. L., CP 66451, México

Jesús Arriaga, Sergio Belmares, Aracelia Alcorta, Francisco Hernández and Ricardo Obregón
Facultad de Ciencias Físico-Matemáticas, México

Ernesto Torres
Facultad de Medicina, México
Universidad Autónoma de Nuevo León, San Nicolás de los Garza, N. L., México

Víctor Coello
CICESE, Monterrey, PIIT, Apodaca, N. L., México

Francisco Paraguay
CIMAV, Chihuahua, Complejo Ind. Chih., Chihuahua, Chih., México

Paulina Segovia
Facultad de Ciencias Físico-Matemáticas, México
CICESE, Monterrey, PIIT, Apodaca, N. L., México

Printed in the USA
CPSIA information can be obtained
at www.ICGtesting.com
JSHW011343221024
72173JS00003B/203

9 781632 382153